Modeling Uncertainties in DC-DC Converters

with MATLAB® and PLECS®

Synthesis Lectures on Electrical Engineering

Editor
Richard C. Dorf, *University of California, Davis*

Modeling Uncertainties in DC-DC Converters with MATLAB® and PLECS®
Farzin Asadi, Sawai Pongswatd, Kei Eguchi, and Ngo Lam Trung
2019

Analytical Solutions for Two Ferromagnetic Nanoparticles Immersed in a Magnetic Field:
Mathematical Model in Bispherical Coordinates
Gehan Anthonys
2018

Circuit Analysis Laboratory Workbook
Teri L. Piatt and Kyle E. Laferty
2017

Understanding Circuits: Learning Problem Solving Using Circuit Analysis
Khalid Sayood
2006

Learning Programming Using MATLAB
Khalid Sayood
2006

Modeling Uncertainties in DC-DC Converters with MATLAB® and PLECS®

Farzin Asadi, Sawai Pongswatd, Kei Eguchi, and Ngo Lam Trung

ISBN: 978-3-031-00892-4   paperback
ISBN: 978-3-031-02020-9   ebook
ISBN: 978-3-031-00135-2   hardcover

DOI 10.1007/978-3-031-02020-9

A Publication in the Springer series
SYNTHESIS LECTURES ON ELECTRICAL ENGINEERING

Lecture #6
Series Editor: Richard C. Dorf, *University of California, Davis*
Series ISSN
Print 1559-811X   Electronic 1559-8128

Modeling Uncertainties in DC-DC Converters

with MATLAB® and PLECS®

Farzin Asadi
Kocaeli University, Kocaeli, Turkey

Sawai Pongswatd
King Mongkut's Institute of Technology Ladkrabang, Bangkok, Thailand

Kei Eguchi
Fukuoka Institute of Technology, Fukuoka, Japan

Ngo Lam Trung
Hanoi University of Science and Technology, Hanoi, Vietnam

SYNTHESIS LECTURES ON ELECTRICAL ENGINEERING #6

ABSTRACT

Modeling is the process of formulating a mathematical description of the system. A model, no matter how detailed, is never a completely accurate representation of a real physical system. A mathematical model is always just an approximation of the true, physical reality of the system dynamics.

Uncertainty refers to the differences or errors between model and real systems and whatever methodology is used to present these errors will be called an uncertainty model. Successful robust control-system design would depend on, to a certain extent, an appropriate description of the perturbation considered.

Modeling the uncertainties in the switch mode DC-DC converters is an important step in designing robust controllers. This book studies different techniques which can be used to extract the uncertain model of DC-DC converters. Once the uncertain model is extracted, robust control techniques such as H_∞ and μ synthesis can be used to design the robust controller.

The book composed of two case studies. The first one is a buck converter and the second one is a Zeta converter. MATLAB® programming is used extensively throughout the book. Some sections use PLECS® as well.

This book is intended to be guide for both academicians and practicing engineers.

KEYWORDS

additive uncertainty, buck converter, DC-DC power conversion, H_∞ control, interval plant, Kharitonov's theorem, multiplicative uncertainty, robust analysis, robust control, state space averaging, uncertainty, uncertainty models, unstructured uncertainty, Zeta converter.

Dedicated to our parents and our lovely families.

Contents

Acknowledgments xi

1   Modeling Uncertainties for a Buck Converter 1

1.1   Introduction 1

1.2   Uncertainty Model 2

    1.2.1   Parametric Uncertainty 2

    1.2.2   Unstructured Uncertainty 3

    1.2.3   Structured Uncertainty 3

1.3   Robust Control 4

    1.3.1   Kharitonov's Theorem 4

    1.3.2 H_∞ Control 4

    1.3.3 μ Synthesis 5

1.4   Dynamics of a Buck Converter Without Uncertainty 6

1.5   Effect of Component Variations 18

1.6   Obtaining the Unstructured Uncertainty Model of the Converter 34

1.7   Obtaining the Interval Plant Model of the Converter 46

1.8   Obtaining the Unstructured Uncertainty Model of the Converter Using
PLECS® 53

1.9   Conclusion 58

    References 59

2   Modeling Uncertainties for a Zeta Converter 63

2.1   Introduction 63

2.2   The Zeta Converter 63

2.3   Calculation of Steady-State Operating Point of the Converter 65

2.4   Drawing the Voltage Gain Ratio 78

2.5   Obtaining the Small Signal Transfer Functions of Converter 81

2.6   Effect of Load Changes on the Small Signal Transfer Functions 92

2.7   Extraction of Additive/Multiplicative Uncertainty Models 102

2.8   Upper Bound of Additive/Multiplicative Uncertainty Models 114

    2.8.1   Extraction of Uncertainty Weights Using the Manual Method 118

2.8.2   Extraction of Uncertainty Weights Using the MATLAB®Ucover
Command 131

2.9   Testing the Obtained Uncertainty Weights 153

2.10   Effect of Components Tolerances 175

2.11   Obtaining the Uncertain Model of the Converter in Precence of
Components Tolerances 187

2.12   Testing the Obtained Uncertainty Weights 201

2.13   Calculating the Maximum/Minimum of the Transfer Function Coefficients 216

2.14   Analyzing the System Without Uncertainty 234

2.15   Audio Susceptibility 249

2.16   Output Impedance 249

2.17   Using the PLECS® to Extract the Uncertain Model of the DC-DC
Converters 257

2.17.1 Additive Uncertainty Model 257

2.17.2 Multiplicative Uncertainty Model 274

2.18   Conclusion 278

Authors' Biographies **279**

Acknowledgments

The authors gratefully acknowledge the MathWorks® and Plexim® support for this project.

Farzin Asadi, Sawai Pongswatd, Kei Eguchi, and Ngo Lam Trung
November 2018

CHAPTER 1

Modeling Uncertainties for a Buck Converter

1.1   INTRODUCTION

A switched DC-DC converter takes the voltage from DC source and converts the voltage of supply into another DC voltage level. They are used to increase or decrease the input voltage level.

The switched DC-DC converters have become an essential component of industrial applications over the past decades. Their high efficiency, small size, low weight, and reduced cost make them a good alternative for conventional linear power supplies, even at low power levels.

Switched DC-DC converters are nonlinear variable structure systems. Various techniques can be found in literature to obtain a Linear Time Invariant (LTI) model of a switched DC-DC converter. The most well-known methods are: current injected approach [1], circuit averaging [2], and State Space Averaging (SSA) [3]. A comprehensive survey of the modeling issues can be found in [4].

Dynamics of switched DC-DC converters change under different output load conditions and/or input voltage. Components tolerances affect the converter dynamics as well. So, a switched DC-DC converter can be modeled as an uncertain dynamical system. Robust control techniques can be used to design controllers for such systems. Robust control refers to control synthesis, its performance and stability, not only to the nominal plant model, but also to the whole family of models in the area of permitted uncertainty of the modeling (perturbations are bounded). So, new terms appear such as robust performance (guaranteed performance of the control system for all systems in the uncertainty region) and robust stability (stability of all possible systems inside the uncertainty modeling bounds).

A number of robust control techniques have been applied successfully to DC-DC converters. Obtaining the uncertain model of the converter is an important step to design a robust controller. This book is a tutorial on modeling the uncertainties in DC-DC converters. Although the main theme of the book is modeling the uncertainties in DC-DC converters, it can be used as a tutorial on implementation of SSA in MATLAB® environment as well.

We studied a buck converter as an illustrative example in this chapter to show different modeling techniques. However, the studied techniques are general and can be applied to other types of switching DC-DC converters.

This chapter is organized as follows. Different types of uncertainty are studied in the Section 1.2. Section 1.3 reviews some of the important robust controller design techniques applied to DC-DC converters as well. Dynamics of a buck converter without uncertainty are studied in Section 1.4. Effect of changes in components values and operating conditions on the converter dynamics is studied in Section 1.5. Extraction of unstructured additive/multiplicative uncertainty model and interval transfer function model is studied in Sections 1.6 and 1.7, respectively. Usefulness of PLECS®in extraction of unstructured uncertainty models of DC-DC converters is studied in Section 1.8. Finally, suitable conclusions are drawn in Section 1.9.

1.2   UNCERTAINTY MODEL

Modeling is the process of formulating a mathematical description of the system. A model, no matter how detailed, is never a completely accurate representation of a real physical system. A mathematical model is always just an approximation of the true physical reality of the system dynamics.

Uncertainty refers to the differences or errors between model and real systems and whatever methodology is used to present these errors will be called an uncertainty model. Successful robust control-system design would depend on, to a certain extent, an appropriate description of the perturbation considered.

1.2.1   PARAMETRIC UNCERTAINTY

Inaccurate description of component characteristics, torn-and-worn effects on plant components, or shifting of operating points cause dynamic perturbations in many industrial control systems. Such perturbations may be represented by variations of certain system parameters over some possible value ranges. They affect the low-frequency range performance and are called "parametric uncertainties." Studying an example is quite helpful. Assume the simple RLC circuit shown in Fig. 1.1.

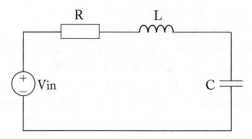

Figure 1.1: Typical RLC circuit.

The transfer function between the capacitor voltage and the input voltage can be written as:

$$\frac{v_c(s)}{v_{\text{in}}(s)} = \frac{\frac{1}{LC}}{s^2 + \frac{R}{L}s + \frac{1}{LC}} = \frac{a}{s^2 + bs + a},$$

(1.1)

where $a = \frac{1}{LC}$ and $b = \frac{R}{L}$. R, L, and C can be written as $R = R_0 + \delta_R$, $L = L_0 + \delta_L$ and $C = C_0 + \delta_C \cdot \delta_R$, δ_L, and δ_C show the effect of aging, measurement error, replacement of the components, etc.

a and b can be written in the same way as, $a_0 + \delta_a$ and $b_0 + \delta_b$, respectively. $a_0 = \frac{1}{L_0 C_0}$ and $b_0 = \frac{R_0}{L_0}$ show the nominal values of a and b, respectively. According to the values of R_0, L_0, C_0, δ_R, δ_L, and δ_C, a, and b can be written as:

$$a_{\min} < a < a_{\max}$$
$$b_{\min} < b < b_{\max}.$$

So, Equation 1.1 no longer describes a single transfer function. It is a family of transfer functions with uncertain coefficients. It has parametric uncertainty.

1.2.2   UNSTRUCTURED UNCERTAINTY

Many dynamic perturbations that may occur in different parts of a system can, however, be lumped into one single perturbation block Δ, for instance, some unmodeled, high-frequency dynamics. This uncertainty representation is referred to as "unstructured" uncertainty. In the case of linear, time-invariant systems, the block Δ may be represented by an unknown transfer function matrix. The unstructured dynamics uncertainty in a control system can be described in different ways [5]. The most famous ones are additive and input/output multiplicative perturbation configurations. If $G_p(s)$, $G_o(s)$, and I show the perturbed system dynamics, a nominal model description of the physical system and identity matrix, respectively, then:

- additive perturbation: $G_p(s) = G_o(s) + \Delta(s)$;

- input multiplicative perturbation: $G_p(s) = G_o(s) \times [I + \Delta(s)]$; and

- output multiplicative perturbation: $G_p(s) = [I + \Delta(s)] \times G_o(s)$.

In Single-Input Single-Output (SISO) systems, there is no difference between Input multiplicative perturbation and output multiplicative perturbation. In Multi-Input Multi-Output (MIMO) systems the two descriptions are not necessarily the same.

1.2.3   STRUCTURED UNCERTAINTY

In some problems the uncertain parts can be taken out from the dynamics and the whole system can be rearranged in a standard configuration of (upper) Linear Fractional Transformation

$F(M, \Delta)$. The uncertian block Δ would then have the following general form:

$$\Delta = \text{diag} \{\delta_1 I_{r_1}, \delta_2 I_{r_2}, \delta_3 I_{r_3}, \ldots, \delta_s I_{r_s}, \Delta_1, \ldots, \Delta_f\}, \qquad \delta_i \in \mathbb{C}, \Delta_j \in \mathbb{C}^{m_j \times m_j},$$

where $\sum_{i=1}^{s} r_i + \sum_{j=1}^{f} m_j = n$ where n is the dimension of the block Δ. So, Δ consist of s repeated scalar blocks and f full blocks. The full blocks need not be square. Since the Δ considered has a certain structure, such description is called structured.

1.3   ROBUST CONTROL

Robust control is a design methodology that explicitly deals with uncertainty. Robust control designs a controller such that:

- some level of performance of the controlled system is guaranteed; and

- irrespective of the changes in the plant dynamics/process dynamics within a predefined class the stability is guaranteed.

Some of the well-known robust control design techniques are studied briefly below.

1.3.1   KHARITONOV'S THEOREM

Kharitonov's theorem is used to assess the stability of a dynamical system when the physical parameters of the system are uncertain. It can be considered as a generalization of Routh–Hurwitz stability test. Routh–Hurwitz is concerned with an ordinary polynomial, i.e., a polynomial with fixed coefficients, while Kharitonov's theorem can study the stability of polynomials with uncertain (varying) coefficients.

Kharitonov's theorem is an analysis tool more than a synthesis tool. Kharitonov's theorem can be used to tune simple controllers such as PID. Designing high-order kontrollers using Kharitonov's theorem is not so common.

[6] and [7] are good references for control engineering applications of Kharitonov's thorem. Plenty of tools and related theorems are gathered there.

Kharitonov's theorem used to design robust controller for DC-DC converters. For instance, [8] used Kharitonov's theorem to tune the PI controller of a quadratic buck converter. [9] designed a robust lead-lag controller for a buck converter.

1.3.2 H_∞ CONTROL

H_∞ techniques formulates the required design specifications (control goles) as an optimal control problem in the frequency domain. In order to do this, some fictitious weighting functions are added to the system model. Weighting functions are selected with respect to the required design specifications. Selection of weights is not a trivial task and require some trial and error to obtain the desired specifications. In fact, the most crucial and difficult task in robust controller design is a choice of the weighting functions. [10], [11], and [12] give very general guidelines for selection of the weights. [13] and [14] used intelligent optimization methods (genetic algorithm) to find the best weighting functions.

The H_∞ design does not always ensure robust stability and robust performance of the closed loop system. This is the main disadvantage of H_∞ design techniques.

H_∞ techniques are studied in many papers and books. Some of the well-known references are introduced here. [15] is the pioneering work of Zames which introduced the H_∞ control. [16] is a good tutorial paper on H_∞ control with some numeric examples. [17] and [18] are general texts on robust control and studied the H_∞ control in detail. [5] and [19] are good references to learn how to design H_∞ controllers using MATLAB®.

H_∞ techniques are applied to a number of DC-DC converters successfully. [20] is the main antecedent in the use of H_∞ to DC-DC converters. It designed a H_∞ controller for a boost converter. Output impedance, audio susceptibility, phase margin, and bandwidth of the control loop are the usual measure of performance in the DC-DC switch mode PWM converters. Reduction of output impedance in nonminimum phase converters (such as boost and buck-boost) is achieved at the expense of phase-margin reduction. However, the H_∞ controller designed in [20] minimizes the output impedance in a wide frequency range without decreasing the phase margin. This paper neglegted the system uncertainties. [21] studied the robust control of boost converters in presence of uncertainties. [22] designed a H_∞ for a High Gain Boost Converter (HGBC).

[23] designed a H_∞ controller to maximize the band width of the control loop with a perfect tracking of the desired output voltage for boost and buck-boost converters. Experimental results are compared with those obtained using Sliding Mode Control (SMC) and current peak control.

[24] used the H_∞ loopshaping to design a controller for a buck-boost converter. Designed controller showed better performance in comparison with PID controller.

[25] and [26] and designed H_∞ controller for paralleled buck converter operating in current-mode control (CMC) and voltage-mode control(VMC).

1.3.3 μ SYNTHESIS

The μ synthesis uses the D-K or μ-K iteration methods [5] to minimize the peak value of the structured singular value of the closed-loop transfer function matrix over the set of all stabilizing controllers K. The structured singular value of a closed-loop system transfer matrix $M(s)$, with uncertainty Δ and singular values σ is defined as:

$$\|M\|_\mu = \mu_\Delta^{-1}(M) \overset{\min}{\Delta \in \Delta} \{\overline{\sigma}(\Delta) : \det(I - M\Delta) = 0\}.$$

Usually the controller designed using the μ synthesis has a high order which makes the implementation difficult. A model order reduction procedure is usually required.

The H_∞ control design techniques, consider the system uncertinty in the unstructured form so the controller designed using the H_∞ techniques is conservative. The μ synthesis considers the uncertainty structures so its output is less conservative [27].

μ synthesis is used to design controller for a number of DC-DC converters successfully. [28] modeled the parametric uncertainties of a buck-boost converter as unstructured and designed the controller using the μ synthesis.

[29] designed a controller for a Quasi Resonant Converters (QRC) operating in Continuous Current Mode (CCM) using the μ synthesis. Robust control of a parallel resonant converter based on the μ synthesis is studied in [30].

[31] gives the general guidelines for designing the robust controller for switch mode DC-DC converters using the μ synthesis.

Table 1.1 compares the different types of robust controllers [27].

Table 1.1: Comparison between linear robust controllers

Linear Robust Controller	Advantages	Disadvantages
Kharitonov's controller	• Simplicity of method • Lower degree of controller	• Direct reduction of output impedance is impossible
H_∞ controller	• Relatively low degree of controller • Direct reduction of output impedance is possible	• Difficulty in determination of weighting functions • Synthesis conservatism
μ controller	• Non conservative • Direct reduction of output impedance is possible • Stability robustness under a wider range of load variations	• Difficulty in determination of weighting functions • High degree of controller • Larger settling times

1.4   DYNAMICS OF A BUCK CONVERTER WITHOUT UNCERTAINTY

Schematic of a buck converter is shown in Fig. 1.2. The buck converter composed of two switches: a MOSFET switch and a diode. In this schematic, Vg, rg, L, rL, C, rC, and R show input DC source, input DC source internal resistance, inductor, inductor Equivalent Series Resistance (ESR), capacitor, capacitor ESR and load, respectively. iO is a fictitious current source added to the schematic in order to calculate the output impedance of converter. In this section we assume that converter works in Continuous Current Mode (CCM). MOSFET switch is controlled with the aid of a Pulse Width Modulator (PWM) controller. MOSFET switch keeps closed for D.T seconds and $(1 - D).T$ seconds open. D and T show duty ratio and switching period, respectively.

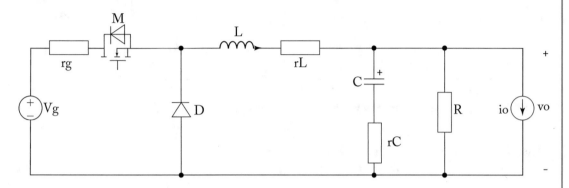

Figure 1.2: Schematic of a buck converter.

When MOSFET is closed, the diode is opened (Fig. 1.3).

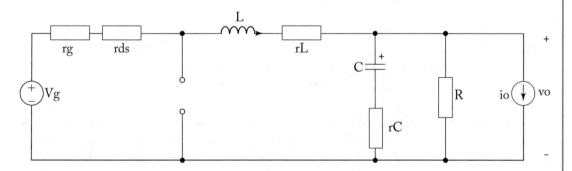

Figure 1.3: Equivalent circuit of a buck converter for closed MOSFET.

The circuit differential equations can be written as:

$$\begin{cases} \dfrac{di_L(t)}{dt} = \dfrac{1}{L}\left(-\left(r_g + r_{ds} + r_L + \dfrac{R \times r_C}{R + r_C}\right)i_L - \dfrac{R}{R + r_C}v_C + \dfrac{R \times r_C}{R + r_C}i_O + v_g\right) \\[4mm] \dfrac{dv_C(t)}{dt} = \dfrac{1}{C}\left(\dfrac{R}{R + r_C}i_L - \dfrac{1}{R + r_C}v_C - \dfrac{R}{R + r_C}i_O\right) \end{cases}$$

$$v_o = r_C C \dfrac{dv_C}{dt} + v_C = \dfrac{R \times r_C}{R + r_C}i_L + \dfrac{R}{R + r_C}v_C - \dfrac{R \times r_C}{R + r_C}i_O.$$

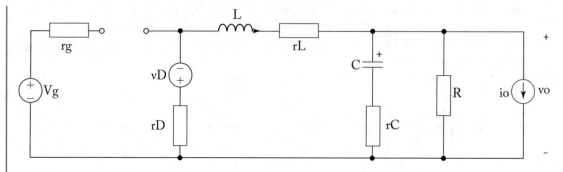

Figure 1.4: Equivalent circuit of a buck converter for opened MOSFET.

When MOSFET is opened, the diode is closed (Fig. 1.4).

The circuit differential equations can be written as:

$$\begin{cases} \dfrac{di_L(t)}{dt} = \dfrac{1}{L}\left(-\left(r_D + r_L + \dfrac{R \times r_C}{R + r_C}\right)i_L - \dfrac{R}{R + r_C}v_C + \dfrac{R \times r_C}{R + r_C}i_O - v_D\right) \\ \dfrac{dv_C(t)}{dt} = \dfrac{1}{C}\left(\dfrac{R}{R + r_C}i_L - \dfrac{1}{R + r_C}v_C - \dfrac{R}{R + r_C}i_O\right) \end{cases}$$

$$v_o = r_C C \dfrac{dv_C}{dt} + v_C = \dfrac{R \times r_C}{R + r_C}i_L + \dfrac{R}{R + r_C}v_C - \dfrac{R \times r_C}{R + r_C}i_O.$$

State Space Averaging (SSA) can be used to extract the small signal transfer functions of the DC-DC converter. The procedure of state space averaging is explained in detail in [32] and [33].

SSA has two important steps: averaging and linearizing the equivalent circuits dynamical equations. Doing the SSA procedure manually is tedious and error prone. A program can be quite useful for this purpose. MATLAB® can be quite helpful for this purpose. The following program shows the implementation of SSA for a converter with component values, as shown in Table 1.2. The component values are assumed to be certain, i.e., have no uncertainty.

After running the following code, the results shown in Figs. 1.5, 1.6, and 1.7 are obtained.

Table 1.2: The buck converter parameters (see Fig. 1.2)

	Nominal Value
Output voltage, Vo	20 V
Duty ratio, D	0.4
Input DC source voltage, Vg	50 V
Input DC source internal resistance, rg	0.05 Ω
MOSFET Drain-Source resistance, rds	40 mΩ
Capacitor, C	100 μF
Capacitor Equivaluent Series Resistance (ESR), rC	0.05 Ω
Inductor, L	400 μH
Inductor ESR, rL	10 mΩ
Diode voltage drop, vD	0.7 V
Diode forward resistance, rD	10 mΩ
Load resistor, R	20 Ω
Switching Frequency, Fsw	20 KHz

```
%This program extracts the small signal transfer functions
%of a Buck converter
clc
clear all

%converter components values
%fsw= 20 KHz
VG=50;      %input DC source voltage
rg=0.5;     %input DC source internal resistance
rds=0.04;   %MOSFET drain-source resistance
rD=0.01;    %Diode series resistance
VD=0.7;     %Diode voltage drop
rL=10e-3;   %Inductor Equivalent Series Resistance(ESR)
L=400e-6;   %Inductor value
rC=0.05;    %Capacitor ESR
C=100e-6;   %Capacitor value
R=20;       %Load resistor
D=0.4;      %Duty ratio
```

```
IO=0;        %Average value of output current source

syms iL vC io vg vD d
% iL : Inductor L1 current
% vC : Capacitor C1 voltage
% io : Output current source
% vg : Input DC source
% vD : Diode voltage drop
% d  : Duty cycle

%Closed MOSFET Equations
diL_dt_MOSFET_close=(-(rg+rds+rL+R*rC/(R+rC))*iL-R/(R+rC)*vC+R*rC
   /(R+rC)*io+vg)/L;
dvC_dt_MOSFET_close=(R/(R+rC)*iL-1/(R+rC)*vC-R/(R+rC)*io)/C;
vo_MOSFET_close=R*rC/(R+rC)*iL+R/(R+rC)*vC-R*rC/(R+rC)*io;

%Opened MOSFET Equations
diL_dt_MOSFET_open=(-(rD+rL+rC*R/(R+rC))*iL-R/(R+rC)*vC+R*rC
   /(R+rC)*io-vD)/L;
dvC_dt_MOSFET_open=(R/(R+rC)*iL-1/(R+rC)*vC-R/(R+rC)*io)/C;
vo_MOSFET_open=R*rC/(R+rC)*iL+R/(R+rC)*vC-R*rC/(R+rC)*io;

%Averaging
averaged_diL_dt=simplify(d*diL_dt_MOSFET_close+(1-d)*
   diL_dt_MOSFET_open);
averaged_dvC_dt=simplify(d*dvC_dt_MOSFET_close+(1-d)*
   dvC_dt_MOSFET_open);
averaged_vo=simplify(d*vo_MOSFET_close+(1-d)*vo_MOSFET_open);

%Substituting the steady values of: input DC voltage source,
%Diode voltage drop, Duty cycle and output current source
%and calculating the DC operating point(IL and VC)
right_side_of_averaged_diL_dt=subs(averaged_diL_dt,[vg vD d io],
   [VG VD D IO]);
right_side_of_averaged_dvC_dt=subs(averaged_dvC_dt,[vg vD d io],
   [VG VD D IO]);

DC_OPERATING_POINT=
solve(right_side_of_averaged_diL_dt==0,
```

```
    right_side_of_averaged_dvC_dt==0,'iL','vC');

IL=eval(DC_OPERATING_POINT.iL);
VC=eval(DC_OPERATING_POINT.vC);
VO=eval(subs(averaged_vo,[iL vC io],[IL VC IO]));

disp('Operating point of converter')
disp('---------------------------')
disp('IL(A)=')
disp(IL)
disp('VC(V)=')
disp(VC)
disp('VO(V)=')
disp(VO)
disp('---------------------------')

%Linearizing the averaged equations around the DC operating
%point. We want to obtain the matrix A, B, C and D
%        .
%        x=Ax+Bu
%        y=Cx+Du
%
%where,
%        x=[iL vC]'
%        u=[io vg d]'
%since we used the variables D for steady state
%duty ratio and C to show the capacitors values
%we use AA, BB, CC, and DD instead of A, B, C, and D.

%Calculating the matrix A
A11=subs(simplify(diff(averaged_diL_dt,iL)),[iL vC d io],
    [IL VC D IO]);
A12=subs(simplify(diff(averaged_diL_dt,vC)),[iL vC d io],
    [IL VC D IO]);

A21=subs(simplify(diff(averaged_dvC_dt,iL)),[iL vC d io],
    [IL VC D IO]);
A22=subs(simplify(diff(averaged_dvC_dt,vC)),[iL vC d io],
    [IL VC D IO]);
```

```
AA=eval([A11 A12;
         A21 A22]);

%Calculating the matrix B
B11=subs(simplify(diff(averaged_diL_dt,io)),[iL vC d vD io vg],
    [IL VC D VD IO VG]);
B12=subs(simplify(diff(averaged_diL_dt,vg)),[iL vC d vD io vg],
    [IL VC D VD IO VG]);
B13=subs(simplify(diff(averaged_diL_dt,d)),[iL  vC d vD io vg],
    [IL VC D VD IO VG]);

B21=subs(simplify(diff(averaged_dvC_dt,io)),[iL vC d vD io vg],
    [IL VC D VD IO VG]);
B22=subs(simplify(diff(averaged_dvC_dt,vg)),[iL vC d vD io vg],
    [IL VC D VD IO VG]);
B23=subs(simplify(diff(averaged_dvC_dt,d)),[iL  vC d vD io vg],
    [IL VC D VD IO VG]);

BB=eval([B11 B12 B13;
         B21 B22 B23]);

%Calculating the matrix C
C11=subs(simplify(diff(averaged_vo,iL)),[iL vC d io],
    [IL VC D IO]);
C12=subs(simplify(diff(averaged_vo,vC)),[iL vC d io],
    [IL VC D IO]);

CC=eval([C11 C12]);

D11=subs(simplify(diff(averaged_vo,io)),[iL vC d vD io vg],
    [IL VC D VD IO VG]);
D12=subs(simplify(diff(averaged_vo,vg)),[iL vC d vD io vg],
    [IL VC D VD IO VG]);
D13=subs(simplify(diff(averaged_vo,d)),[iL  vC d vD io vg],
    [IL VC D VD IO VG]);

%Calculating the matrix D
```

```
DD=eval([D11 D12 D13]);

%Producing the State Space Model and obtaining
%the small signal transfer functions
sys=ss(AA,BB,CC,DD);
sys.inputname={'io';'vg';'d'};
sys.outputname={'vo'};

vo_io=tf(sys(1,1)); %Output impedance transfer function
                    %vo(s)/io(s)
vo_vg=tf(sys(1,2)); %vo(s)/vg(s)
vo_d=tf(sys(1,3));  %Control-to-output(vo(s)/d(s))

%drawing the Bode diagrams
figure(1)
bode(vo_io),grid minor,title('vo(s)/io(s)')

figure(2)
bode(vo_vg),grid minor,title('vo(s)/vg(s)')

figure(3)
bode(vo_d),grid minor,title('vo(s)/d(s)')
```

According to the analysis results, the converter with parameters given in Table 1.2 can be modeled, as shown in Fig. 1.8.

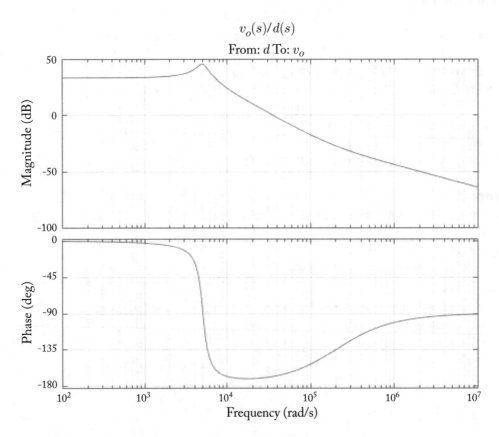

Figure 1.5: $\frac{v_o(s)}{d(s)} = 6257.7\frac{s+2\times10^5}{s^2+1203s+2.523\times10^7}$ Bode diagram.

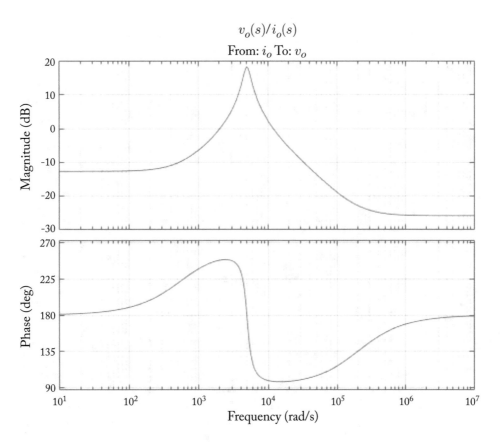

Figure 1.6: $\frac{v_o(s)}{i_o(s)} = -0.0499\frac{(s+2\times10^5)(s+580)}{s^2+1203s+2.523\times10^7}$ Bode diagram.

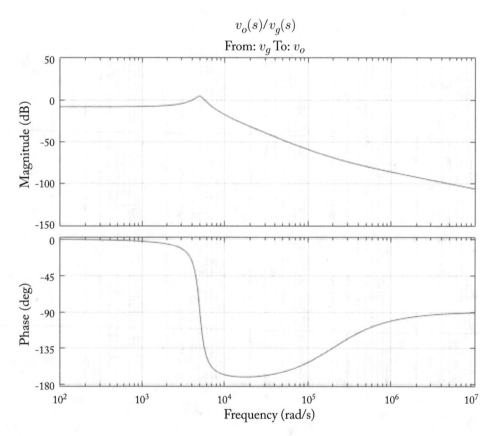

Figure 1.7: $\frac{v_o(s)}{v_g(s)} = 49.875 \frac{(s+2\times10^5)}{s^2+1203s+2.523\times10^7}$ Bode diagram.

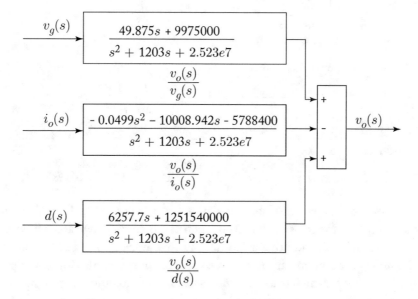

Figure 1.8: Dynamic model of the converter.

1.5   EFFECT OF COMPONENT VARIATIONS

Assume a buck converter composed of uncertain components. The components nominal values and their variation is given in Table 1.3.

Table 1.3: The buck converter parameters

	Nominal Value	Variations
Output voltage, Vo	20 V	0%
Input DC source voltage, Vg	50 V	±20%
Input DC source internal resistance, rg	0.5 Ω	±20%
MOSFET Drain-Source resistance, rds	40 mΩ	±20%
Capacitor, C	100 μF	±20%
Capacitor Equivaluent Series Resistance (ESR), rC	0.05 Ω	-10%, +90%
Inductor, L	400 μH	±10%
Inductor ESR, rL	10 mΩ	-10%, +90%
Diode voltage drop, vD	0.7 V	±30%
Diode forward resistance, rD	10 mΩ	-10%, +50%
Load resistor, R	20 Ω	±20%

We want to study the effect of these variations on the converter dynamics. Figures 1.9–1.11 show the changes in the transfer functions when parameters change according to Table 1.3.

We want to obtain the additive and multiplicative uncertainty description of the buck converter. The previous program can be modified for this purpose. The following program defines the buck converter parameters as uncertain values. Each iteration of program samples the uncertain set and applies the SSA to sampled values. Calculated transfer functions are compared with their nominal values (result of previous analysis) to obtain the additive and multiplicative description of the uncertainty.

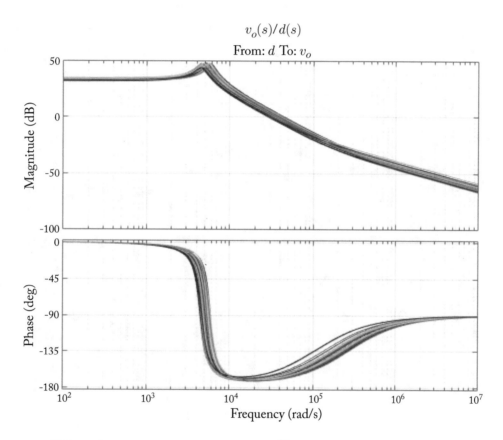

Figure 1.9: Effect of components changes on the $\frac{v_o(s)}{d(s)}$.

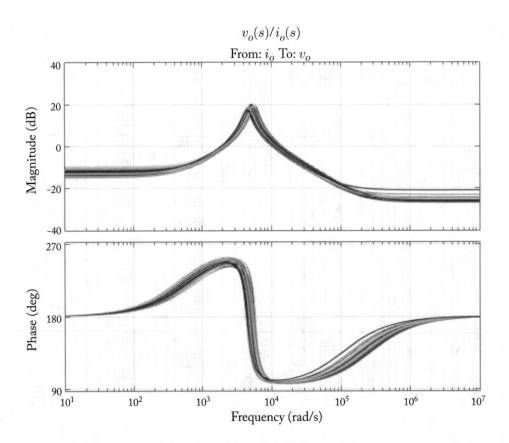

Figure 1.10: Effect of components changes on the $\frac{v_o(s)}{i_o(s)}$.

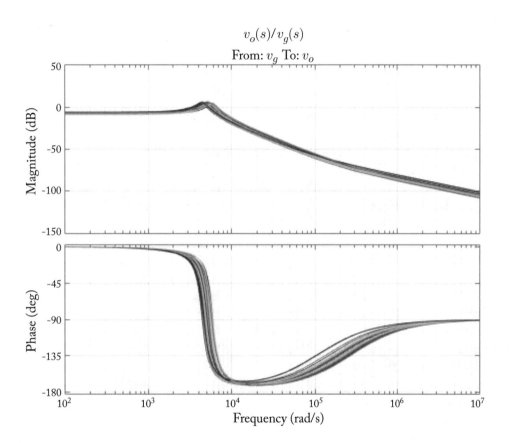

Figure 1.11: Effect of components changes on the $\frac{v_o(s)}{v_g(s)}$.

```
%This program calculates the small signal transfer functions of
%Buck converter and draws the multiplicative and additive
%uncertainty due to variations in parameters.

clc
clear all

NumberOfIteration=50;
DesiredOutputVoltage=20;

s=tf('s');
vo_io_nominal=-0.049875*(s+2e5)*(s+580)/(s^2+1203*s+2.523e7);
    %Nominal vo(s)/io(s)
vo_vg_nominal=49.875*(s+2e5)/(s^2+1203*s+2.523e7);
    %Nominal vo(s)/vg(s)
vo_d_nominal=6257.7*(s+2e5)/(s^2+1203*s+2.523e7);
    %Nominal vo(s)/d(s)

n=0;
for i=1:NumberOfIteration
n=n+1;
%Definition of uncertainity in parameters
VG_unc=ureal('VG_unc',50,'Percentage',[-20 +20]);
    %Average value of input DC source is in the range of 40..60
rg_unc=ureal('rds_unc',.5,'Percentage',[-20 +20]);
    %Input DC source resistance
rds_unc=ureal('rds_unc',.04,'Percentage',[-20 +20]);
    %MOSFET on resistance
C_unc=ureal('C_unc',100e-6,'Percentage',[-20 +20]);
    %Capacitor value
rC_unc=ureal('rC_unc',.05,'Percentage',[-10 +90]);
    %Capacitor Equivalent Series Resistance(ESR)
L_unc=ureal('L_unc',400e-6,'Percentage',[-10 +10]);
    %Inductor value
rL_unc=ureal('rL_unc',0.01,'Percentage',[-10 +90]);
    %Inductor Equivalent Series Resistance(ESR)
rD_unc=ureal('rD_unc',.01,'Percentage',[-10 +50]);
    %Diode series resistance
```

```
VD_unc=ureal('VD_unc',.7,'Percentage',[-30 +30]);
   %Diode voltage drop
R_unc=ureal('R_unc',20,'Percentage',[-20 +20]);
   %Load resistance
IO=0;
   %Average value of output current source

%Sampling the uncertain set for instance usample(VG_unc,1)
%takes one sample of uncertain parameter VG_unc

VG=usample(VG_unc,1);      %Sampled average value of input DC
                           %source
rg=usample(rg_unc,1);      %Sampled internal resistance of input
                           %DC source
rds=usample(rds_unc,1);    %Sampled MOSFET on resistance
C=usample(C_unc,1);        %Sampled capacitor value
rC=usample(rC_unc,1);      %Sampled capacitor Equivalent Series
                           %Resistance(ESR)
L=usample(L_unc,1);        %Sampled inductor value
rL=usample(rL_unc,1);      %Sampled inductor Equivalent Series
                           %Resistance(ESR)
rD=usample(rD_unc,1);      %Sampled diode series resistance
VD=usample(VD_unc,1);      %Sampled diode voltage drop
R=usample(R_unc,1);        %Sampled load resistance

%output voltage of an IDEAL(i.e., no losses) Buck converter
%operating in CCM is given by:
%VO=D.VG
%where
%VO: average value of output voltage
%D: Duty Ratio
%VG: Input DC voltage
%So, for a IDEAL converter
%       VO
%D=----------
%       VG
%Since our converter has losses we use a bigger duty ratio,
%for instance:
%               VO
```

```
%D=1.05 --------
%               VG

D=1.05*DesiredOutputVoltage/(VG);        %Duty cylcle

syms iL vC io vg vD d
% iL : Inductor L1 current
% vC : Capacitor C1 voltage
% io : Output current source
% vg : Input DC source
% vD : Diode voltage drop
% d  : Duty cycle

%Closed MOSFET Equations
diL_dt_MOSFET_close=(-(rg+rds+rL+R*rC/(R+rC))*iL-R/(R+rC)*vC+R*rC
    /(R+rC)*io+vg)/L;
dvC_dt_MOSFET_close=(R/(R+rC)*iL-1/(R+rC)*vC-R/(R+rC)*io)/C;
vo_MOSFET_close=R*rC/(R+rC)*iL+R/(R+rC)*vC-R*rC/(R+rC)*io;

%Opened MOSFET Equations
diL_dt_MOSFET_open=(-(rD+rL+rC*R/(R+rC))*iL-R/(R+rC)*vC+R*rC
    /(R+rC)*io-vD)/L;
dvC_dt_MOSFET_open=(R/(R+rC)*iL-1/(R+rC)*vC-R/(R+rC)*io)/C;
vo_MOSFET_open=R*rC/(R+rC)*iL+R/(R+rC)*vC-R*rC/(R+rC)*io;

%Averaging
averaged_diL_dt=simplify(d*diL_dt_MOSFET_close+(1-d)*
    diL_dt_MOSFET_open);
averaged_dvC_dt=simplify(d*dvC_dt_MOSFET_close+(1-d)*
    dvC_dt_MOSFET_open);
averaged_vo=simplify(d*vo_MOSFET_close+(1-d)*vo_MOSFET_open);

%Substituting the steady values of: input DC voltage source,
%Diode voltage drop, Duty cycle and output current source and
%calculating the DC operating point(IL and VC)
right_side_of_averaged_diL_dt=subs(averaged_diL_dt,[vg vD d io],
    [VG VD D IO]);
right_side_of_averaged_dvC_dt=subs(averaged_dvC_dt,[vg vD d io],
    [VG VD D IO]);
```

```
DC_OPERATING_POINT=
solve(right_side_of_averaged_diL_dt==0,
   right_side_of_averaged_dvC_dt==0,'iL','vC');

IL=eval(DC_OPERATING_POINT.iL);
VC=eval(DC_OPERATING_POINT.vC);
VO=eval(subs(averaged_vo,[iL vC io],[IL VC IO]));

disp('Operating point of converter')
disp('----------------------------')
disp('IL(A)=')
disp(IL)
disp('VC(V)=')
disp(VC)
disp('VO(V)=')
disp(VO)
disp('----------------------------')

%Linearizing the averaged equations around the DC operating
%point. We want to obtain the matrix A, B, C, and D
%          .
%        x=Ax+Bu
%        y=Cx+Du
%
%where,
%        x=[iL1 iL2 vC1 vC2]'
%        u=[io vg d]'
%since we used the variables D for steady state duty ratio
%and C to show the capacitors values we use AA, BB, CC and
%DD instead of A, B, C, and D.

%Calculating the matrix A
A11=subs(simplify(diff(averaged_diL_dt,iL)),[iL vC d io],
   [IL VC D IO]);
A12=subs(simplify(diff(averaged_diL_dt,vC)),[iL vC d io],
   [IL VC D IO]);

A21=subs(simplify(diff(averaged_dvC_dt,iL)),[iL vC d io],
```

```
      [IL VC D IO]);
A22=subs(simplify(diff(averaged_dvC_dt,vC)),[iL vC d io],
      [IL VC D IO]);



AA=eval([A11 A12;
          A21 A22]);

%Calculating the matrix B
B11=subs(simplify(diff(averaged_diL_dt,io)),[iL vC d vD io vg],
      [IL VC D VD IO VG]);
B12=subs(simplify(diff(averaged_diL_dt,vg)),[iL vC d vD io vg],
      [IL VC D VD IO VG]);
B13=subs(simplify(diff(averaged_diL_dt,d)),[iL  vC d vD io vg],
      [IL VC D VD IO VG]);

B21=subs(simplify(diff(averaged_dvC_dt,io)),[iL vC d vD io vg],
      [IL VC D VD IO VG]);
B22=subs(simplify(diff(averaged_dvC_dt,vg)),[iL vC d vD io vg],
      [IL VC D VD IO VG]);
B23=subs(simplify(diff(averaged_dvC_dt,d)),[iL  vC d vD io vg],
      [IL VC D VD IO VG]);

BB=eval([B11 B12 B13;
          B21 B22 B23]);

%Calculating the matrix C
C11=subs(simplify(diff(averaged_vo,iL)),[iL vC d io],
      [IL VC D IO]);
C12=subs(simplify(diff(averaged_vo,vC)),[iL vC d io],
      [IL VC D IO]);

CC=eval([C11 C12]);

D11=subs(simplify(diff(averaged_vo,io)),[iL vC d vD io vg],
      [IL VC D VD IO VG]);
D12=subs(simplify(diff(averaged_vo,vg)),[iL vC d vD io vg],
      [IL VC D VD IO VG]);
D13=subs(simplify(diff(averaged_vo,d)),[iL  vC d vD io vg],
```

```
   [IL VC D VD IO VG]);

%Calculating the matrix D
DD=eval([D11 D12 D13]);

%Producing the State Space Model and obtaining the small
%signal transfer functions
sys=ss(AA,BB,CC,DD);
sys.inputname={'io';'vg';'d'};
sys.outputname={'vo'};

vo_io=tf(sys(1,1)); %Output impedance transfer function
                    %vo(s)/io(s)
vo_vg=tf(sys(1,2)); %vo(s)/vg(s)
vo_d=tf(sys(1,3));  %Control-to-output(vo(s)/d(s))

%drawing the Bode diagrams
%Multiplicative uncertainty
figure(1)
bodemag((vo_io-vo_io_nominal)/vo_io_nominal),grid minor,
   title('Multiplicative uncertainty in vo(s)/io(s)')
hold on

figure(2)
bodemag((vo_vg-vo_vg_nominal)/vo_vg_nominal),grid minor,
   title('Multiplicative uncertainty in vo(s)/vg(s)')
hold on

figure(3)
bodemag((vo_d-vo_d_nominal)/vo_d_nominal),grid minor,
   title('Multiplicative uncertainty in vo(s)/d(s)')
hold on

%Additive uncertainty
figure(4)
bodemag((vo_io-vo_io_nominal)),grid minor,title('Additive
   uncertainty in vo(s)/io(s)')
hold on
```

```
figure(5)
bodemag((vo_vg-vo_vg_nominal)),grid minor,title('Additive
    uncertainty in vo(s)/vg(s)')
hold on

figure(6)
bodemag((vo_d-vo_d_nominal)),grid minor,title('Additive
    uncertainty in vo(s)/d(s)')
hold on

disp('Percentage of work done:')
disp(n/NumberOfIteration*100) %shows the progress of the loop
disp('')
end
```

After running the code, results shown in Figs. 1.12–1.17 are obtained.

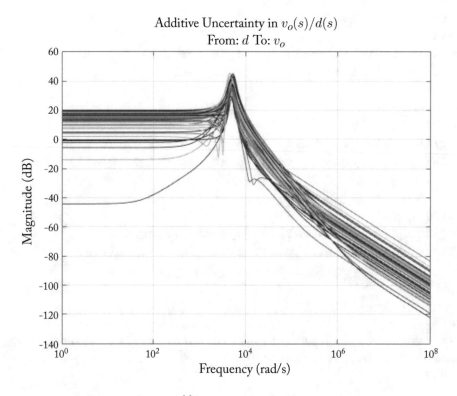

Figure 1.12: Additive uncertainty in $\frac{v_o(s)}{d(s)}$.

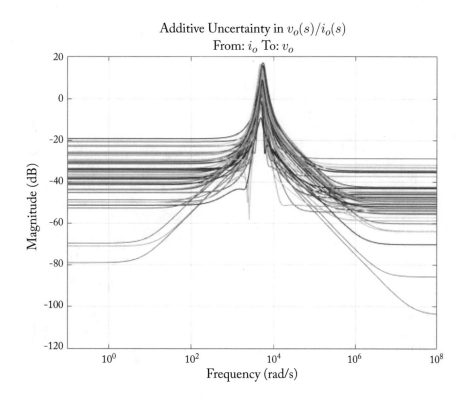

Figure 1.13: Additive uncertainty in $\frac{v_o(s)}{i_o(s)}$.

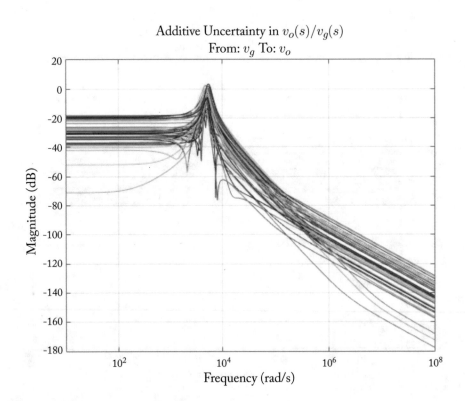

Figure 1.14: Additive uncertainty in $\frac{v_o(s)}{v_g(s)}$.

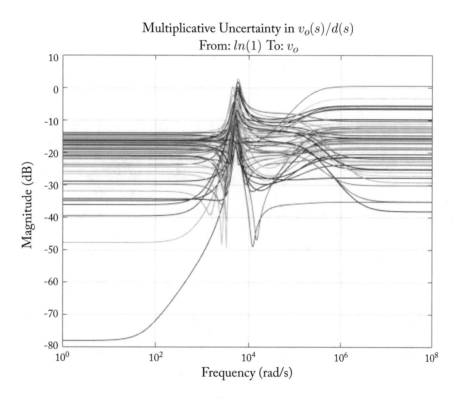

Figure 1.15: Multiplicative uncertainty in $\frac{v_o(s)}{d(s)}$.

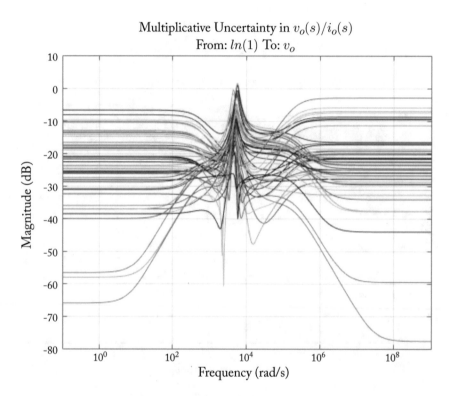

Figure 1.16: Multiplicative uncertainty in $\frac{v_o(s)}{i_o(s)}$.

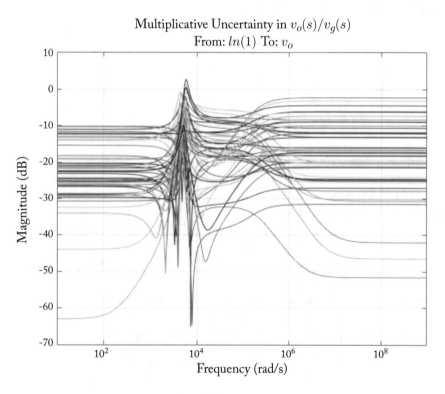

Figure 1.17: Multiplicative uncertainty in $\frac{v_o(s)}{v_g(s)}$.

1.6 OBTAINING THE UNSTRUCTURED UNCERTAINTY MODEL OF THE CONVERTER

Although the result of previous analysis shows the effect of variations on the converters dynamics graphically, it gives no information about the suitable weights to model the uncertainty. MATLAB® has a function named "ucover" which can be used to obtain the suitable weights to cover the uncertainty. The previous program is modified to extract the uncertain weights. The following program extract the multiplicative uncertainty weights. Additive uncertainty weights can be extracted in the same way if required.

```
%This program calculates the small signal transfer
%functions of Buck converter and extracts the
%multiplicative uncertainty weights. We use ucover
%command in this program to find the weights

clc
clear all

NumberOfIteration=50;
DesiredOutputVoltage=20;

s=tf('s');
vo_io_nominal=-0.049875*(s+2e5)*(s+580)/(s^2+1203*s+2.523e7);
    %Nominal vo(s)/io(s)
vo_vg_nominal=49.875*(s+2e5)/(s^2+1203*s+2.523e7);
    %Nominal vo(s)/vg(s)
vo_d_nominal=6257.7*(s+2e5)/(s^2+1203*s+2.523e7);
    %Nominal vo(s)/d(s)

n=0;
for i=1:NumberOfIteration
n=n+1;
%Definition of uncertainty in parameters
VG_unc=ureal('VG_unc',50,'Percentage',[-20 +20]);
    %Average value of input DC source is in the range of 40..60
rg_unc=ureal('rds_unc',.5,'Percentage',[-20 +20]);
    %Input DC source resistance
rds_unc=ureal('rds_unc',.04,'Percentage',[-20 +20]);
    %MOSFET on resistance
C_unc=ureal('C_unc',100e-6,'Percentage',[-20 +20]);
```

```
   %Capacitor value
rC_unc=ureal('rC_unc',.05,'Percentage',[-10 +90]);
   %Capacitor Equivalent Series Resistance(ESR)
L_unc=ureal('L_unc',400e-6,'Percentage',[-10 +10]);
   %Inductor value
rL_unc=ureal('rL_unc',0.01,'Percentage',[-10 +90]);
   %Inductor Equivalent Series Resistance(ESR)
rD_unc=ureal('rD_unc',.01,'Percentage',[-10 +50]);
   %Diode series resistance
VD_unc=ureal('VD_unc',.7,'Percentage',[-30 +30]);
   %Diode voltage drop
R_unc=ureal('R_unc',20,'Percentage',[-20 +20]);
   %Load resistance
IO=0;
   %Average value of output current source

%Sampling the uncertain set
%for instance usample(VG_unc,1) takes one sample of uncertain
%parameter VG_unc

VG=usample(VG_unc,1);
   %Sampled average value of input DC source
rg=usample(rg_unc,1);
   %Sampled internal resistance of input DC source
rds=usample(rds_unc,1);
   %Sampled MOSFET on resistance
C=usample(C_unc,1);
   %Sampled capacitor value
rC=usample(rC_unc,1);
   %Sampled capacitor Equivalent Series Resistance(ESR)
L=usample(L_unc,1);
   %Sampled inductor  value
rL=usample(rL_unc,1);
   %Sampled inductor  Equivalent Series Resistance(ESR)
rD=usample(rD_unc,1);
   %Sampled diode series resistance
VD=usample(VD_unc,1);
   %Sampled diode voltage drop
R=usample(R_unc,1);
```

```
    %Sampled load resistance

%output voltage of an IDEAL(i.e., no losses) Buck converter
%operating in CCM is given by:
%VO=D.VG
%where
%VO: average value of output voltage
%D: Duty Ratio
%VG: Input DC voltage
%So, for a IDEAL converter
%       VO
%D=----------
%       VG
%Since our converter has losses we use a bigger duty ratio,
%for instance:
%            VO
%D=1.05 --------
%            VG

D=1.05*DesiredOutputVoltage/(VG);       %Duty cylcle

syms iL vC io vg vD d
%iL : Inductor L1 current
%vC : Capacitor C1 voltage
%io : Output current source
%vg : Input DC source
%vD : Diode voltage drop
%d  : Duty cycle

%Closed MOSFET Equations
diL_dt_MOSFET_close=(-(rg+rds+rL+R*rC/(R+rC))*iL-R/(R+rC)*
    vC+R*rC/(R+rC)*io+vg)/L;
dvC_dt_MOSFET_close=(R/(R+rC)*iL-1/(R+rC)*vC-R/(R+rC)*io)/C;
vo_MOSFET_close=R*rC/(R+rC)*iL+R/(R+rC)*vC-R*rC/(R+rC)*io;

%Opened MOSFET Equations
diL_dt_MOSFET_open=(-(rD+rL+rC*R/(R+rC))*iL-R/(R+rC)*
    vC+R*rC/(R+rC)*io-vD)/L;
dvC_dt_MOSFET_open=(R/(R+rC)*iL-1/(R+rC)*vC-R/(R+rC)*io)/C;
```

```
vo_MOSFET_open=R*rC/(R+rC)*iL+R/(R+rC)*vC-R*rC/(R+rC)*io;

%Averaging
averaged_diL_dt=simplify(d*diL_dt_MOSFET_close+(1-d)*
    diL_dt_MOSFET_open);
averaged_dvC_dt=simplify(d*dvC_dt_MOSFET_close+(1-d)*
    dvC_dt_MOSFET_open);
averaged_vo=simplify(d*vo_MOSFET_close+(1-d)*vo_MOSFET_open);

%Substituting the steady values of: input DC voltage source,
%Diode voltage drop, Duty cycle and output current source
%and calculating the DC operating point(IL and VC)
right_side_of_averaged_diL_dt=subs(averaged_diL_dt,[vg vD d io],
    [VG VD D IO]);
right_side_of_averaged_dvC_dt=subs(averaged_dvC_dt,[vg vD d io],
    [VG VD D IO]);

DC_OPERATING_POINT=
solve(right_side_of_averaged_diL_dt==0,
    right_side_of_averaged_dvC_dt==0,'iL','vC');

IL=eval(DC_OPERATING_POINT.iL);
VC=eval(DC_OPERATING_POINT.vC);
VO=eval(subs(averaged_vo,[iL vC io],[IL VC IO]));

disp('Operating point of converter')
disp('---------------------------')
disp('IL(A)=')
disp(IL)
disp('VC(V)=')
disp(VC)
disp('VO(V)=')
disp(VO)
disp('---------------------------')

%Linearizing the averaged equations around the DC operating
%point. We want to obtain the matrix A, B, C, and D
%         .
%    x=Ax+Bu
```

```
%        y=Cx+Du
%
%where,
%      x=[iL vC]'
%      u=[io vg d]'
%since we used the variables D for steady state duty
%ratio and C to show the capacitors values we use AA,
%BB, CC and DD instead of A, B, C, and D.


%Calculating the matrix A
A11=subs(simplify(diff(averaged_diL_dt,iL)),[iL vC d io],
    [IL VC D IO]);
A12=subs(simplify(diff(averaged_diL_dt,vC)),[iL vC d io],
    [IL VC D IO]);

A21=subs(simplify(diff(averaged_dvC_dt,iL)),[iL vC d io],
    [IL VC D IO]);
A22=subs(simplify(diff(averaged_dvC_dt,vC)),[iL vC d io],
    [IL VC D IO]);


AA=eval([A11 A12;
         A21 A22]);

%Calculating the matrix B
B11=subs(simplify(diff(averaged_diL_dt,io)),[iL vC d vD io vg],
    [IL VC D VD IO VG]);
B12=subs(simplify(diff(averaged_diL_dt,vg)),[iL vC d vD io vg],
    [IL VC D VD IO VG]);
B13=subs(simplify(diff(averaged_diL_dt,d)),[iL  vC d vD io vg],
    [IL VC D VD IO VG]);

B21=subs(simplify(diff(averaged_dvC_dt,io)),[iL vC d vD io vg],
    [IL VC D VD IO VG]);
B22=subs(simplify(diff(averaged_dvC_dt,vg)),[iL vC d vD io vg],
    [IL VC D VD IO VG]);
B23=subs(simplify(diff(averaged_dvC_dt,d)),[iL  vC d vD io vg],
    [IL VC D VD IO VG]);
```

```
BB=eval([B11 B12 B13;
         B21 B22 B23]);

%Calculating the matrix C
C11=subs(simplify(diff(averaged_vo,iL)),[iL vC d io],
    [IL VC D IO]);
C12=subs(simplify(diff(averaged_vo,vC)),[iL vC d io],
    [IL VC D IO]);

CC=eval([C11 C12]);

D11=subs(simplify(diff(averaged_vo,io)),[iL vC d vD io vg],
    [IL VC D VD IO VG]);
D12=subs(simplify(diff(averaged_vo,vg)),[iL vC d vD io vg],
    [IL VC D VD IO VG]);
D13=subs(simplify(diff(averaged_vo,d)),[iL  vC d vD io vg],
    [IL VC D VD IO VG]);

%Calculating the matrix D
DD=eval([D11 D12 D13]);

%Producing the State Space Model and obtaining the
%small signal transfer functions
sys=ss(AA,BB,CC,DD);
sys.inputname={'io';'vg';'d'};
sys.outputname={'vo'};

vo_io=tf(sys(1,1)); %Output impedance transfer function
                    %vo(s)/io(s)
vo_vg=tf(sys(1,2)); %vo(s)/vg(s)
vo_d=tf(sys(1,3));  %Control-to-output(vo(s)/d(s))

if n==1
    %there is no variable names "array" in the first running
    %of the loop. variable "array" is initialize in the first
    %running of loop.
    array1=vo_d;
    array2=vo_vg;
    array3=vo_io;
```

```
else
    array1=stack(1,array1,vo_d);
    array2=stack(1,array2,vo_vg);
    array3=stack(1,array3,vo_io);
end
disp('Percentage of work done:')
disp(n/NumberOfIteration*100) %shows the progress of the loop
disp('')
end

%Calculating the multiplicative uncertainity bounds
omega=logspace(-1,5,200);
array1_frd=frd(array1,omega);
array2_frd=frd(array2,omega);
array3_frd=frd(array3,omega);

relerr1 = (vo_d_nominal-array1_frd)/vo_d_nominal;
relerr2 = (vo_vg_nominal-array2_frd)/vo_vg_nominal;
relerr3 = (vo_io_nominal-array3_frd)/vo_io_nominal;

[P1,Info1]=ucover(array1_frd,vo_d_nominal,2);
    %fitting a 2nd order weight to vo(s)/d(s).
[P2,Info2]=ucover(array2_frd,vo_vg_nominal,2);
    %fitting a 2nd order weight to vo(s)/vg(s).
[P3,Info3]=ucover(array3_frd,vo_io_nominal,2);
    %fitting a 2nd order weight to vo(s)/io(s).

%uncertainty weight of vo(s)/d(s)
figure(1)
bodemag(relerr1,'b--',Info1.W1,'r',{0.1,100000});

%uncertainty weight of vo(s)/vg(s)
figure(2)
bodemag(relerr2,'b--',Info2.W1,'r',{0.1,100000});

%uncertainty weight of vo(s)/io(s)
figure(3)
bodemag(relerr3,'b--',Info3.W1,'r',{0.1,100000});
```

After running the code, the analysis results shown in Figs. 1.18, 1.19, and 1.20 are obtained. According to the obtained weights, the diagram shown in Fig. 1.21 can be suggested to model the system. Δ_1, Δ_2, and Δ_3 are stable and unknown transfer functions satisfying the condition $\|\Delta_1\|_\infty \le 1$, $\|\Delta_2\|_\infty \le 1$, and $\|\Delta_3\|_\infty \le 1$.

The block diagram shown in Fig. 1.21 can be transferred to the standard $M - \Delta$ representation (see Fig. 1.22) using simple block diagram manipulations.

In Fig. 1.22, $d = \begin{bmatrix} d_1 \\ d_2 \\ d_3 \end{bmatrix}$, $u = \begin{bmatrix} \widetilde{v}_g(s) \\ \widetilde{i}_o(s) \\ \widetilde{d}(s) \end{bmatrix}$, $z = \begin{bmatrix} z_1 \\ z_2 \\ z_3 \end{bmatrix}$, $w = \widetilde{v}_o(s)$,

$$\Delta = \begin{bmatrix} \Delta_1(s) & 0 & 0 \\ 0 & \Delta_2(s) & 0 \\ 0 & 0 & \Delta_3(s) \end{bmatrix},$$

and

$$M = \begin{bmatrix} 0 & 0 & 0 & W_1(s) & 0 & 0 \\ 0 & 0 & 0 & 0 & W_2(s) & 0 \\ 0 & 0 & 0 & 0 & 0 & W_3(s) \\ \dfrac{\widetilde{v}_o(s)}{v_g(s)} & \dfrac{\widetilde{v}_o(s)}{i_o(s)} & \dfrac{\widetilde{v}_o(s)}{d(s)} & \dfrac{\widetilde{v}_o(s)}{v_g(s)} & \dfrac{\widetilde{v}_o(s)}{i_o(s)} & \dfrac{\widetilde{v}_o(s)}{d(s)} \end{bmatrix}.$$

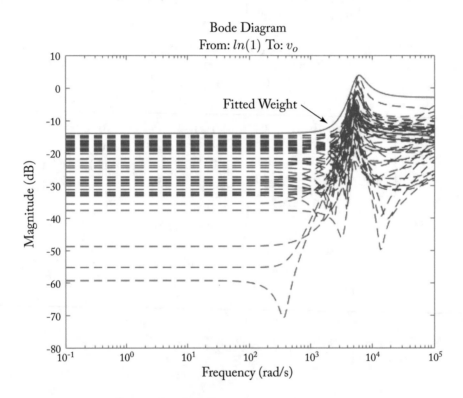

Figure 1.18: $W_1 = 0.7244\frac{s^2+4381s+9.373\times10^6}{s^2+2800s+3.339\times10^7}$ Bode diagram.

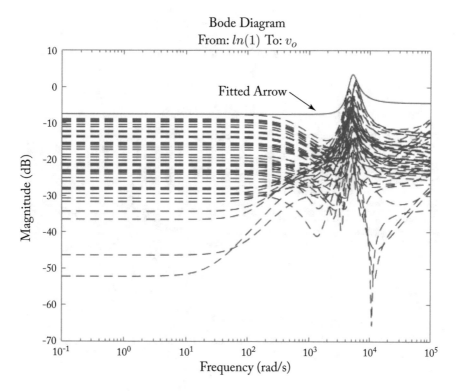

Figure 1.19: $W_2 = 0.752 \frac{s^2+7405s+1.543\times10^7}{s^2+3828s+3.581\times10^7}$ Bode diagram.

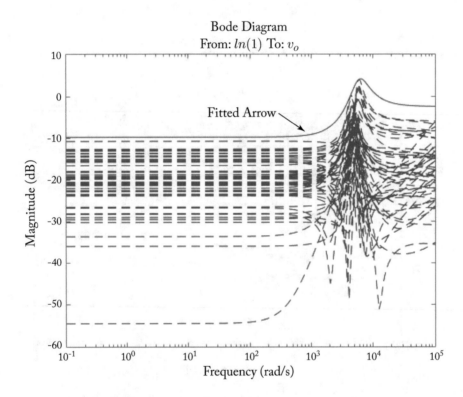

Figure 1.20: $W_3 = 0.609\frac{s^2+3589s+1.817\times10^7}{s^2+1586s+2.598\times10^7}$ Bode diagram.

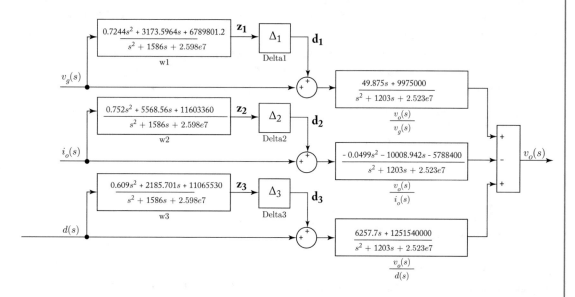

Figure 1.21: Multiplicative uncertainty model of the converter.

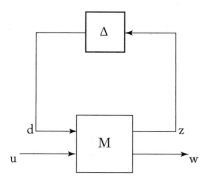

Figure 1.22: $M - \Delta$ model of uncertain systems.

1.7   OBTAINING THE INTERVAL PLANT MODEL OF THE CONVERTER

The model obtained in previous section is suitable for techniques such as H_∞ and μ-synthesis. If you want to design the controller using the Kharitonov's theorem you need the lower/upper bound of transfer function coefficients. The following program extracts the lower/upper bound of transfer function coefficients.

```
%This program calculates the small signal transfer functions
%of Buck converter and extracts the upper/lower bounds for
%transfer function coefficients. This program helps you design
%the controller using Kharitonov's theorem

clc
clear all

NumberOfIteration=150;
DesiredOutputVoltage=20;

s=tf('s');
vo_io_nominal=-0.049875*(s+2e5)*(s+580)/(s^2+1203*s+2.523e7);
    %Nominal vo(s)/io(s)
vo_vg_nominal=49.875*(s+2e5)/(s^2+1203*s+2.523e7);
    %Nominal vo(s)/vg(s)
vo_d_nominal=6257.7*(s+2e5)/(s^2+1203*s+2.523e7);
    %Nominal vo(s)/d(s)

n=0;
for i=1:NumberOfIteration
n=n+1;
%Definition of uncertainity in parameters
VG_unc=ureal('VG_unc',50,'Percentage',[-20 +20]);
    %Average value of input DC source is in the range of 40..60
rg_unc=ureal('rds_unc',.5,'Percentage',[-20 +20]);
    %Input DC source resistance
rds_unc=ureal('rds_unc',.04,'Percentage',[-20 +20]);
    %MOSFET on resistance
C_unc=ureal('C_unc',100e-6,'Percentage',[-20 +20]);
    %Capacitor value
rC_unc=ureal('rC_unc',.05,'Percentage',[-10 +90]);
```

```
%Capacitor Equivalent Series Resistance(ESR)
L_unc=ureal('L_unc',400e-6,'Percentage',[-10 +10]);
   %Inductor value
rL_unc=ureal('rL_unc',0.01,'Percentage',[-10 +90]);
   %Inductor Equivalent Series Resistance(ESR)
rD_unc=ureal('rD_unc',.01,'Percentage',[-10 +50]);
   %Diode series resistance
VD_unc=ureal('VD_unc',.7,'Percentage',[-30 +30]);
   %Diode voltage drop
R_unc=ureal('R_unc',20,'Percentage',[-20 +20]);
   %Load resistance
IO=0;
   %Average value of output current source

%Sampling the uncertain set
%for instance usample(VG_unc,1) takes one sample of uncertain
%parameter VG_unc

VG=usample(VG_unc,1);
   %Sampled average value of input DC source
rg=usample(rg_unc,1);
   %Sampled internal resistance of input DC source
rds=usample(rds_unc,1);
   %Sampled MOSFET on resistance
C=usample(C_unc,1);
   %Sampled capacitor value
rC=usample(rC_unc,1);
   %Sampled capacitor Equivalent Series Resistance(ESR)
L=usample(L_unc,1);
   %Sampled inductor  value
rL=usample(rL_unc,1);
   %Sampled inductor  Equivalent Series Resistance(ESR)
rD=usample(rD_unc,1);
   %Sampled diode series resistance
VD=usample(VD_unc,1);
   %Sampled diode voltage drop
R=usample(R_unc,1);
   %Sampled load resistance
```

```
%output voltage of an IDEAL(i.e. no losses) Buck converter
%operating in CCM is given by:
%VO=D.VG
%where
%VO: average value of output voltage
%D: Duty Ratio
%VG: Input DC voltage
%So, for a IDEAL converter
%        VO
%D=----------
%        VG
%Since our converter has losses we use a bigger duty ratio,
%for instance:
%              VO
%D=1.05 --------
%              VG


D=1.05*DesiredOutputVoltage/(VG);        %Duty cylcle


syms iL vC io vg vD d
% iL : Inductor L1 current
% vC : Capacitor C1 voltage
% io : Output current source
% vg : Input DC source
% vD : Diode voltage drop
% d  : Duty cycle


%Closed MOSFET Equations
diL_dt_MOSFET_close=(-(rg+rds+rL+R*rC/(R+rC))*iL-R/(R+rC)*
   vC+R*rC/(R+rC)*io+vg)/L;
dvC_dt_MOSFET_close=(R/(R+rC)*iL-1/(R+rC)*vC-R/(R+rC)*io)/C;
vo_MOSFET_close=R*rC/(R+rC)*iL+R/(R+rC)*vC-R*rC/(R+rC)*io;


%Opened MOSFET Equations
diL_dt_MOSFET_open=(-(rD+rL+rC*R/(R+rC))*iL-R/(R+rC)*
   vC+R*rC/(R+rC)*io-vD)/L;
dvC_dt_MOSFET_open=(R/(R+rC)*iL-1/(R+rC)*vC-R/(R+rC)*io)/C;
vo_MOSFET_open=R*rC/(R+rC)*iL+R/(R+rC)*vC-R*rC/(R+rC)*io;
```

```
%Averaging
averaged_diL_dt=simplify(d*diL_dt_MOSFET_close+(1-d)*
   diL_dt_MOSFET_open);
averaged_dvC_dt=simplify(d*dvC_dt_MOSFET_close+(1-d)*
   dvC_dt_MOSFET_open);
averaged_vo=simplify(d*vo_MOSFET_close+(1-d)*vo_MOSFET_open);

%Substituting the steady values of: input DC voltage source,
%Diode voltage drop, Duty cycle and output current source and
%calculating the DC operating point(IL and VC)
right_side_of_averaged_diL_dt=subs(averaged_diL_dt,[vg vD d io],
   [VG VD D IO]);
right_side_of_averaged_dvC_dt=subs(averaged_dvC_dt,[vg vD d io],
   [VG VD D IO]);

DC_OPERATING_POINT=
solve(right_side_of_averaged_diL_dt==0,
   right_side_of_averaged_dvC_dt==0,'iL','vC');

IL=eval(DC_OPERATING_POINT.iL);
VC=eval(DC_OPERATING_POINT.vC);
VO=eval(subs(averaged_vo,[iL vC io],[IL VC IO]));

disp('Operating point of converter')
disp('---------------------------')
disp('IL(A)=')
disp(IL)
disp('VC(V)=')
disp(VC)
disp('VO(V)=')
disp(VO)
disp('---------------------------')

%Linearizing the averaged equations around the DC operating
%point. We want to obtain the matrix A, B, C, and D
%          .
%       x=Ax+Bu
%       y=Cx+Du
%
```

```
%where,
%      x=[iL vC]'
%      u=[io vg d]'
%since we used the variables D for steady state duty ratio
%and C to show the capacitors values we use AA, BB, CC, and
%DD instead of A, B, C, and D.

%Calculating the matrix A
A11=subs(simplify(diff(averaged_diL_dt,iL)),[iL vC d io],
    [IL VC D IO]);
A12=subs(simplify(diff(averaged_diL_dt,vC)),[iL vC d io],
    [IL VC D IO]);

A21=subs(simplify(diff(averaged_dvC_dt,iL)),[iL vC d io],
    [IL VC D IO]);
A22=subs(simplify(diff(averaged_dvC_dt,vC)),[iL vC d io],
    [IL VC D IO]);



AA=eval([A11 A12;
         A21 A22]);

%Calculating the matrix B
B11=subs(simplify(diff(averaged_diL_dt,io)),[iL vC d vD io vg],
    [IL VC D VD IO VG]);
B12=subs(simplify(diff(averaged_diL_dt,vg)),[iL vC d vD io vg],
    [IL VC D VD IO VG]);
B13=subs(simplify(diff(averaged_diL_dt,d)),[iL  vC d vD io vg],
    [IL VC D VD IO VG]);

B21=subs(simplify(diff(averaged_dvC_dt,io)),[iL vC d vD io vg],
    [IL VC D VD IO VG]);
B22=subs(simplify(diff(averaged_dvC_dt,vg)),[iL vC d vD io vg],
    [IL VC D VD IO VG]);
B23=subs(simplify(diff(averaged_dvC_dt,d)),[iL  vC d vD io vg],
    [IL VC D VD IO VG]);

BB=eval([B11 B12 B13;
         B21 B22 B23]);
```

```
%Calculating the matrix C
C11=subs(simplify(diff(averaged_vo,iL)),[iL vC d io],
    [IL VC D IO]);
C12=subs(simplify(diff(averaged_vo,vC)),[iL vC d io],
    [IL VC D IO]);

CC=eval([C11 C12]);

D11=subs(simplify(diff(averaged_vo,io)),[iL vC d vD io vg],
    [IL VC D VD IO VG]);
D12=subs(simplify(diff(averaged_vo,vg)),[iL vC d vD io vg],
    [IL VC D VD IO VG]);
D13=subs(simplify(diff(averaged_vo,d)),[iL  vC d vD io vg],
    [IL VC D VD IO VG]);

%Calculating the matrix D
DD=eval([D11 D12 D13]);

%Producing the State Space Model and obtaining the small
%signal transfer functions
sys=ss(AA,BB,CC,DD);
sys.inputname={'io';'vg';'d'};
sys.outputname={'vo'};

vo_io=tf(sys(1,1)); %Output impedance transfer function
                    %vo(s)/io(s)
vo_vg=tf(sys(1,2)); %vo(s)/vg(s)
vo_d=tf(sys(1,3));  %Control-to-output(vo(s)/d(s))

%Extracts the transfer function coefficients
if n==1
      [num_vo_io,den_vo_io]=tfdata(vo_io,'v');
      [num_vo_vg,den_vo_vg]=tfdata(vo_vg,'v');
      [num_vo_d,den_vo_d]=tfdata(vo_d,'v');
else
      [num1,den1]=tfdata(vo_io,'v'); %extracts the numerator and
                                     %denominator of vo(s)/io(s)
      num_vo_io=[num_vo_io;num1];    %numerator of vo(s)/io(s)
```

```matlab
        den_vo_io=[den_vo_io;den1];      %denominator of vo(s)/io(s)

        [num2,den2]=tfdata(vo_vg,'v');
        num_vo_vg=[num_vo_vg;num2];
        den_vo_vg=[den_vo_vg;den2];

        [num3,den3]=tfdata(vo_d,'v');
        num_vo_d=[num_vo_d;num3];
        den_vo_d=[den_vo_d;den3];
end
disp('Percentage of work done:')
disp(n/NumberOfIteration*100) %shows the progress of the loop
disp('')
end
disp('')
disp('vo(s)/d(s)')
disp('maximum of numerator coefficients:')
disp(max(num_vo_d))
disp('minimum of numerator coefficients:')
disp(min(num_vo_d))
disp('')
disp('maximum of denominator coefficients:')
disp(max(den_vo_d))
disp('minimum of denominator coefficients:')
disp(min(den_vo_d))
disp('-------------')
disp('vo(s)/io(s)')
disp('maximum of numerator coefficients:')
disp(max(num_vo_io))
disp('minimum of numerator coefficients:')
disp(min(num_vo_io))
disp('')
disp('maximum of denominator coefficients:')
disp(max(den_vo_io))
disp('minimum of denominator coefficients:')
disp(min(den_vo_io))
disp('-------------')
disp('vo(s)/vg(s)')
disp('maximum of numerator coefficients:')
```

```
disp(max(num_vo_vg))
disp('minimum of numerator coefficients:')
disp(min(num_vo_vg))
disp('')
disp('maximum of denominator coefficients:')
disp(max(den_vo_vg))
disp('minimum of denominator coefficients:')
disp(min(den_vo_vg))
disp('-------------')
```

After running the program, the following results obtained appear in Table 1.4.

Table 1.4: Interval plant model of the transfer functions

$\dfrac{v_o(s)}{d(s)}$	$\dfrac{v_o(s)}{i_o(s)}$	$\dfrac{v_o(s)}{v_g(s)}$
$\dfrac{v_o(s)}{d(s)} = \dfrac{b_1 s + b_0}{s^2 + a_1 s + a_0}$	$\dfrac{v_o(s)}{i_o(s)} = \dfrac{b_2 s^2 + b_1 s + b_0}{s^2 + a_1 s + a_0}$	$\dfrac{v_o(s)}{v_g(s)} = \dfrac{b_1 s + b_0}{s^2 + a_1 s + a_0}$
$4.2793 \times 10^3 < b_1 < 1.1302 \times 10^4$	$-0.0883 < b_2 < -0.045$	$38.4197 < b_1 < 118.2969$
$7.9196 \times 10^8 < b_0 < 1.9482 \times 10^9$	$-1.2407 \times 10^4 < b_1 < -8.3366 \times 10^3$	$7.4277 \times 10^6 < b_0 < 1.6682 \times 10^7$
$1.0076 \times 10^3 < a_1 < 1.6746 \times 10^3$	$-1.0087 \times 10^7 < b_0 < -3.8357 \times 10^6$	$1.0076 \times 10^3 < a_1 v 1.6746 \times 10^3$
$1.9979 \times 10^7 < a_0 < 3.3958 \times 10^7$	$1.0076 \times 10^3 < a_1 < 1.6746 \times 10^3$	$1.9979 \times 10^7 < a_0 < 3.3958 \times 10^7$
	$1.9979 \times 10^7 < a_0 < 3.3958 \times 10^7$	

1.8 OBTAINING THE UNSTRUCTURED UNCERTAINTY MODEL OF THE CONVERTER USING PLECS®

PLECS® can be used to obtain the uncertain model of DC-DC converters. PLECS® can extract the dynamic model of DC-DC converters [34]. Assume we want to extract the additive uncertainty model of the buck converter with component values given in Table 1.3. The schematic shown in Fig. 1.23 can be used for this purpose.

The lower buck converter in Fig. 1.23 has nominal values. The upper buck converter component values are generated with the aid of following program.

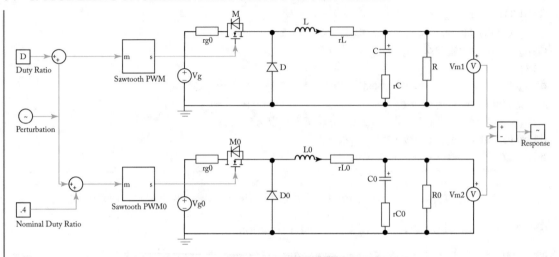

Figure 1.23: PLECS® schematic to extract the additive uncertainty model of the converter.

```
%This program produces random component values
%produced values are copied into the Windows clipboard
%so you can paste them easily into the Initialization
%section of PLECS
clc
clear all

NumberOfIteration=20;      %set the desired number of
                           %iteration here.
DesiredOutputVoltage=20; %set the desired output voltage here.

n=0;
for i=1:NumberOfIteration
n=n+1;
%Definition of uncertainity in parameters
VG_unc=ureal('VG_unc',50,'Percentage',[-20 +20]);
    %Average value of input DC source is in the range of 16..24
rg_unc=ureal('rg_unc',.5,'Percentage',[-20 +20]);
    %Input DC source Internal resistance of input DC source
rds_unc=ureal('rds_unc',.04,'Percentage',[-20 +20]);
    %MOSFET on resistance
C_unc=ureal('C_unc',100e-6,'Percentage',[-20 +20]);
```

```
   %Capacitor C value
rC_unc=ureal('rC_unc',.05,'Percentage',[-10 +90]);
   %Capacitor C Equivalent Series Resistance(ESR)
L_unc=ureal('L_unc',400e-6,'Percentage',[-10 +10]);
   %Inductor L value
rL_unc=ureal('rL_unc',10e-3,'Percentage',[-10 +90]);
   %Inductor L Equivalent Series Resistance(ESR)
rD_unc=ureal('rD_unc',.01,'Percentage',[-10 +50]);
   %Diode series resistance
VD_unc=ureal('VD_unc',.7,'Percentage',[-30 +30]);
   %Diode voltage drop
R_unc=ureal('R_unc',20,'Percentage',[-20 +20]);
   %Load resistance

fsw=20e3;
   %Switching frequency

%Sampling the uncertain set
%for instance usample(VG_unc,1) takes one sample of uncertain
%parameter VG_unc

VG=usample(VG_unc,1);
   %Sampled average value of input DC source
rg=usample(rg_unc,1);
   %Sampled DC source internal resistor
rds=usample(rds_unc,1);
   %Sampled MOSFET on resistance
C=usample(C_unc,1);
   %Sampled capacitor C value
rC=usample(rC_unc,1);
   %Sampled capacitor C Equivalent Series Resistance(ESR)
L=usample(L_unc,1);
   %Sampled inductor L value
rL=usample(rL_unc,1);
   %Sampled inductor L Equivalent Series Resistance(ESR)
rD=usample(rD_unc,1);
   %Sampled diode series resistance
VD=usample(VD_unc,1);
   %Sampled diode voltage drop
```

```
R=usample(R_unc,1);
    %Sampled load resistance

%output voltage of an IDEAL(i.e., no losses) Buck converter
%operating in CCM is given by:
%
%VO=D.VG
%
%where
%VO: average value of output voltage
%D: Duty Ratio
%VG: Input DC voltage
%So, for a IDEAL converter
%        VO
%D=----------
%        VG
%Since our converter has losses we use a bigger duty ratio,
%for instance:
%              VO
%D=1.05 ----------
%              VG
D=1.05*DesiredOutputVoltage/VG;

%preparing the strings
S1=strcat('VG=',num2str(VG),';');
S2=strcat('rg=',num2str(rg),';');
S3=strcat('rds=',num2str(rds),';');
S4=strcat('C=',num2str(C),';');
S5=strcat('rC=',num2str(rC),';');
S6=strcat('L=',num2str(L),';');
S7=strcat('rL=',num2str(rL),';');
S8=strcat('rD=',num2str(rD),';');
S9=strcat('VD=',num2str(VD),';');
S10=strcat('R=',num2str(R),';');
S11=strcat('D=',num2str(D),';');

%coping the data into the Windows Clipboard. So, you can
%paste it into the PLECS
data=strcat(S1,S2,S3,S4,S5,S6,S7,S8,S9,S10,S11);
```

```
clipboard('copy',data)
disp('Data is copied into clipboard. You can paste it in PLECS
initialization section right now...')
message=strcat('Iteration #',num2str(n),' finished.');
disp(message)
disp('Press any key to produce another value set.')
disp(' ')
pause
end
disp('-------------------------')
disp('Program terminates here...')
```

The program randomly select the component values within the acceptable range and copy them into the Windows® clipboard. So, the user can paste it easily into the "Initialization" section (see Fig. 1.24) of PLECS® and run the simulation. PLECS® can export the simulation results as Comma Separated Values (CSV) files so MATLAB® can read the simulation results.

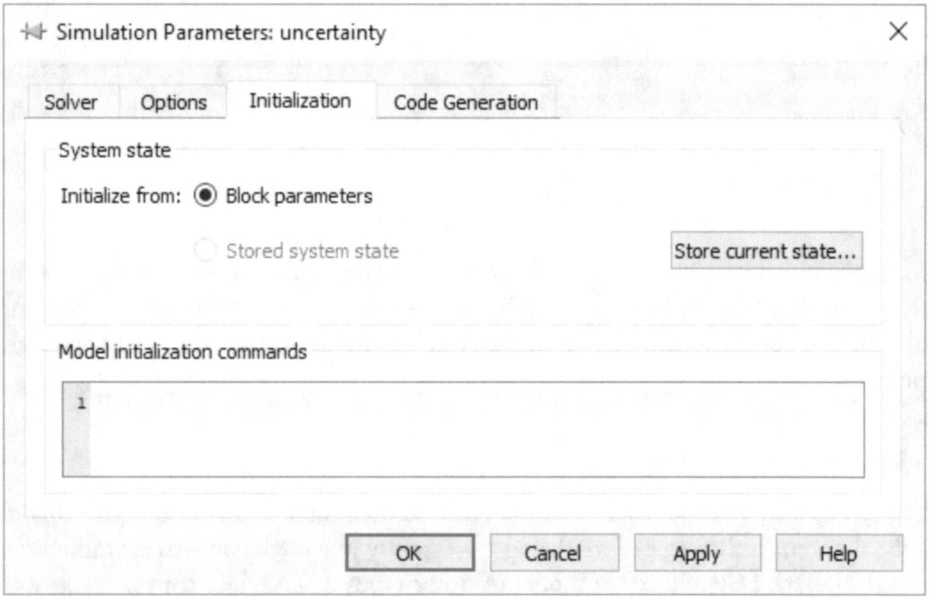

Figure 1.24: Initialization tab of the PLECS®.

Figure 1.25 shows the simulation results for 20 runs of the schematic shown in Fig. 1.23. Upper bound of uncertainty can be extracted using the methods shown before.

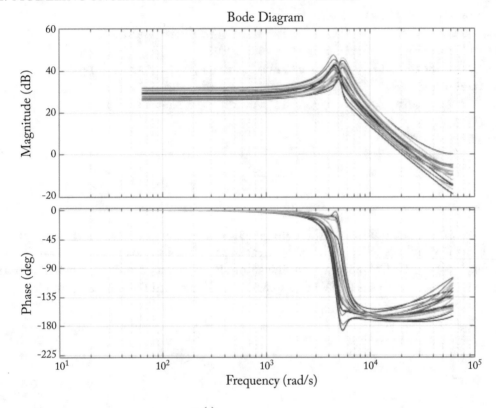

Figure 1.25: Additive uncertainty in $\frac{v_o(s)}{d(s)}$.

Multiplicative uncertainty model of the buck converter can be extracted by dividing the frequency responses (shown in Fig. 1.25) by the nominal frequency response (the one with nominal values of components). Figure 1.26 shows the multiplicative uncertainty model of the buck converter.

1.9   CONCLUSION

Modeling the system uncertainty is an important step of robust control design. This chapter studied the different techniques of modeling uncertainty in a buck converter. Studied methods can be used to extract the uncertain model of other types of DC-DC converters as well. The next chapter studies a Zeta converter which is a fourth-order converter.

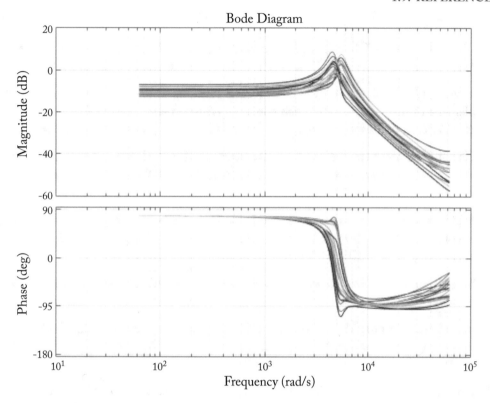

Figure 1.26: Multiplicative uncertainty in $\frac{v_o(s)}{d(s)}$.

REFERENCES

[1] Kislovski, A. S., Ridl, R., and Socal, N. *Dynamic Analysis of Switching Mode DC-DC Converter*, Van Nostrand Reinhold, New York, 1991. DOI: 10.1007/978-94-011-7849-5. 1

[2] Erikson, R. and Maksimovic, D. *Fundamentals of Power Electronics*, Springer, 2007. 1

[3] Middlebrook, R. and Cuk, S. General unified approach to modeling switching-converter power stages. *International Journal of Electronics Theoretical and Experimental* 42(6), pages 521–550, 1977. DOI: 10.1109/pesc.1976.7072895. 1

[4] Maksimovic, D., Stankovic, A., Tottuvelil, V., et al. Modeling and simulation of power electronic converters. *Proc. IEEE*, pages 898–912, 2001. DOI: 10.1109/5.931486. 1

[5] Gu, D., Petkov, P., and Konstantinov, M. *Robust Control Design with MATLAB*, Springer, 2013. DOI: 10.1007/978-1-4471-4682-7. 3, 5

[6] Barmish, R. *javascript:void(0)*, Macmillan, 1993. 4

[7] Bhattacharyya, S., Chapellat, H., and Keel, L. *Robust Control the Parametric Approach*, Prentice Hall PTR, 1995. DOI: 10.1016/b978-0-08-042230-5.50016-5. 4

[8] Bevrani, H., Babahajyani, P., Habibi, F., and Hiyama, T. Robust control design and implementation for a quadratic buck converter. *International Power Electronics Conference ECCE ASIA*, pages 99–103, Sapporo, 2010. DOI: 10.1109/ipec.2010.5543644. 4

[9] Chang, C. Robust control of DC-DC converters: The buck converter. *Power Electronics Specialists Conference*, pages 1094–1097, Atlanta, 1995. DOI: 10.1109/pesc.1995.474951. 4

[10] Lundstron, P., Skogestad, S., and Wang, Z., Performance weight selection for H_∞ and μ-control method. *Transactions on Institute and Control*, 24, pages 1–252, 1991. DOI: 10.1177/014233129101300504. 5

[11] Skogesttad, S. and Postlethwaite, I. *Multivariable Feedback Control-Analysis and Design*, John Wiley & Sons, 2000. 5

[12] Beaven, R., Wright, M., and Seaward, D. Weighting function selection in the H_∞ design process. *Control Engineering Practice*, pages 625–633, 1996. DOI: 10.1016/0967-0661(96)00044-5. 5

[13] Donha, D. and Katebi, M. Automatic weight selection for H_∞ controller synthesis. *International Journal of Systems Science*, pages 651–664, 2007. DOI: 10.1080/00207720701500559. 5

[14] Alfaro-Cid, E., McGookin, E., and Murray-Smith, D. Optimisation of the weighting functions of an H_∞ controller using genetic algorithms and structured genetic algorithms. *International Journal of Systems Science*, pages 335–347, 2008. DOI: 10.1080/00207720701777959. 5

[15] Zames, G. Feedback and optimal sensitivity: Model reference transformations, multiplicative seminorms, and approximate inverses. *IEEE Transactions on Automatic Control*, pages 301–320, 1981. DOI: 10.1109/tac.1981.1102603. 5

[16] Kwakernak, H. Robust control and H_∞ optimization. *Automatica*, pages 255–273, 1993. DOI: 10.1016/0005-1098(93)90122-A. 5

[17] Zhou, K. and Doyle, J. *Essential of Robust Control*, Pearson, 1997. 5

[18] Green, M. and Limbeer, D. *Linear Robust Control*, Dover Publications, 2012. 5

[19] Chiang, R., Safonov, M., Balas, G., and Packard, A. *Robust Control Toolbox*, 3rd ed., The Mathworks, Inc., 2007. 5

[20] Naim, R., Weiss, G., and Ben-Yaakov, S. H_∞ control applied to boost power converters. *IEEE Transactions on Power Electron*, pages 677–683, 1997. DOI: 10.1109/63.602563. 5

[21] Khayat, Y., Naderi, M., Shafiee, Q., et al. Robust control of a DC-DC boost converter: H_2 and H_∞ techniques. *8th Power Electronics, Drive Systems and Technologies Conference (PEDSTC)*, pages 407–412, 2017. 5

[22] Shaw, P. and Veerachary, M. Mixed-sensitivity based robust H_∞ controller design for high-gain boost converter. *International Conference on Computer, Communications and Electronics (Comptelix)*, pages 612–617, 2017. DOI: 10.1109/comptelix.2017.8004042. 5

[23] Vidal-Idiarte, E., Martinez-Salamero, L., Valderrama-Blavi, H., Guinjoan, F., and Maixe, J. Analysis and design of H_∞ control of nonminimum phase-switching converters. *IEEE Transactions on Circuits and Systems I: Fundamental Theory and Applications*, pages 1316–1323, 2003. DOI: 10.1109/tcsi.2003.816337. 5

[24] Hernandez, W. Robust control applied to improve the performance of a buck-boost converter. *WSEAS Transaction on Circuit and Systems*, pages 450–459, 2008. 5

[25] Gadoura, I., Suntio, T., and Zenger, K. Dynamic system modeling and analysis for multi loop operation of paralleled DC/DC converters. *Proc. of the International Conference on Power Electronics and Intelligent Motion*, pages 443–448, 2001. 5

[26] Gadoura, I., Suntio, T., and Zenger, K. Model uncertainty and robust control of paralleled DC/DC converters. *International Conference on Power Electronics, Machines and Drives*, pages 74–79, 2002. DOI: 10.1049/cp:20020092. 5

[27] Bevrani, H., Abrishamchian, M., and Safari-Shad, N. Nonlinear and linear robust control of switching power converters. *Proc. of the IEEE International Conference on Control Applications*, pages 808–813, 1999. DOI: 10.1109/cca.1999.807765. 6

[28] Buso, S. Synthesis of a robust voltage controller for a buck–boost converter. *Proc. IEEE Power Electronics Specialists Conference (PESC)*, pages 766–772, 1996. DOI: 10.1109/pesc.1996.548668. 6

[29] Bevrani, H., Ise, T., Mitani, Y., et al. A robust approach to controller design for DC-DC quasi–resonant converter. *IEEJ Transaction on Industry Application*, pages 91–100, 2004. DOI: 10.1541/ieejias.124.91. 6

[30] Bu, J., Sznaier, M., Wang, Z., et al. Robust controller design for a parallel resonant converter using μ synthesis. *IEEE Transaction on Power Electronics*, pages 837–853, 1997. DOI: 10.1109/63.623002. 6

[31] Wallis, G. and Tymerski, R. Generalized approach for μ-synthesis of robust switching regulators. *IEEE Transactions on Aerospace Electronic Systems*, pages 422–431, 2000. DOI: 10.1109/7.845219. 6

[32] Suntio, T. *Dynamic Profile of Switched Mode Converters: Modeling, Analysis and Control*, Wiley WCH, 2009. DOI: 10.1002/9783527626014. 8

[33] Asadi, F. and Eguchi, K. *Dynamics and Control of DC-DC Converters*, Morgan & Claypool, 2018. DOI: 10.2200/s00828ed1v01y201802pel010. 8

[34] PLECS User Manual. Plexim, 2018. 53

CHAPTER 2

Modeling Uncertainties for a Zeta Converter

2.1   INTRODUCTION

A buck converter which is a second-order converter is studied in Chapter 1. This chapter studies a Zeta converter which is a fourth-order converter.

2.2   THE ZETA CONVERTER

Schematic of a Zeta converter is shown in Fig. 2.1. The Zeta converter composed of two switches: a MOSFET switch and a diode. In this schematic, Vg, rg, L_i, rL_i, C_i, rC_i, and R shows the input DC source, the input DC source internal resistance, ith inductor, ith inductor Equivalent Series Resistance (ESR), ith capacitor, ith capacitor ESR, and load, respectively. iO is a fictitious current source added to the schematic in order to calculate the output impedance of converter. In this section we assume that converter works in Continuous Current Mode (CCM). MOSFET switch is controlled with the aid of a Pulse Width Modulator (PWM) controller. MOSFET switch keeps closed for D.T seconds and $(1 - D).T$ seconds open. D and T show duty ratio and switching period, respectively.

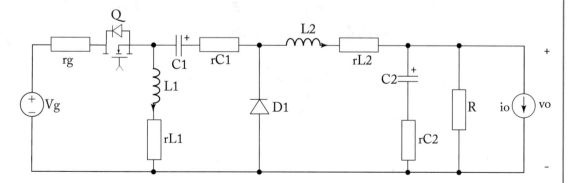

Figure 2.1: Schematic of a Zeta converter.

When MOSFET is closed, the diode is opened (Fig. 2.2).

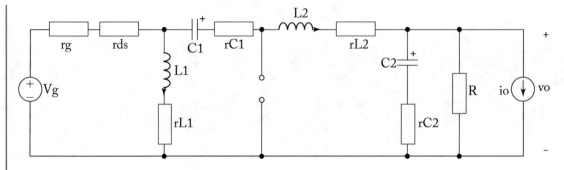

Figure 2.2: Equivalent circuit of a Zeta converter for closed MOSFET.

The circuit differential equations can be written as:

$$
\begin{cases}
L_1 \dfrac{di_{L_1}}{dt} = -\left(r_{L_1} + r_g + r_{ds}\right) i_{L_1} - \left(r_g + r_{ds}\right) i_{L_2} + v_g \\[2mm]
L_2 \dfrac{di_{L_2}}{dt} = -\left(r_g + r_{ds}\right) i_{L_1} - \left(r_g + r_{ds} + r_{C_1} + r_{L_2} + \dfrac{R \times r_{C_2}}{R + r_{C_2}}\right) i_{L_2} \\[2mm]
\qquad\qquad + v_{C_1} - \dfrac{R}{R + r_{C_2}} v_{C_2} + \dfrac{R \times r_{C_2}}{R + r_{C_2}} i_o + v_g \\[2mm]
C_1 \dfrac{dv_{C_1}}{dt} = -i_{L_2} \\[2mm]
C_2 \dfrac{dv_{C_2}}{dt} = \dfrac{R}{R + r_{C_2}} i_{L_2} - \dfrac{1}{R + r_{C_2}} v_{C_2} - \dfrac{R}{R + r_{C_2}} i_o
\end{cases}
$$

$$
v_o = r_{C_2} C_2 \frac{dv_{C_2}}{dt} + v_{C_2} = \frac{R \times r_{C_2}}{R + r_{C_2}} i_{L_2} + \frac{R}{R + r_{C_2}} v_{C_2} - \frac{R \times r_{C_2}}{R + r_{C_2}} i_o.
$$

When MOSFET is opened, the diode is closed (Fig. 2.3).

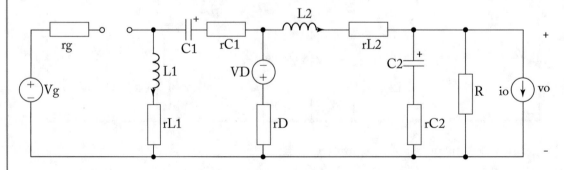

Figure 2.3: Equivalent circuit of a Zeta converter for opened MOSFET.

The circuit differential equations can be written as:

$$\begin{cases} L_1\dfrac{di_{L_1}}{dt} = -\left(r_{L_1} + r_{C_1} + r_D\right)i_{L_1} - r_D i_{L_2} - v_{C_1} - v_D \\ L_2\dfrac{di_{L_2}}{dt} = -r_D i_{L_1} - \left(r_D + r_{L_2} + \dfrac{R \times r_{C_2}}{R + r_{C_2}}\right)i_{L_2} - \dfrac{R}{R + r_{C_2}}v_{C_2} + \dfrac{R \times r_{C_2}}{R + r_{C_2}}i_o - v_D \\ C_1\dfrac{dv_{C_1}}{dt} = i_{L_1} \\ C_2\dfrac{dv_{C_2}}{dt} = \dfrac{R}{R + r_{C_2}}i_{L_2} - \dfrac{1}{R + r_{C_2}}v_{C_2} - \dfrac{R}{R + r_{C_2}}i_o \end{cases}$$

$$v_o = r_{C_2}C_2\frac{dv_{C_2}}{dt} + v_{C_2} = \frac{R \times r_{C_2}}{R + r_{C_2}}i_{L_2} + \frac{R}{R + r_{C_2}}v_{C_2} - \frac{R \times r_{C_2}}{R + r_{C_2}}i_o.$$

2.3 CALCULATION OF STEADY-STATE OPERATING POINT OF THE CONVERTER

Consider a Zeta converter with the component values, as shown in Table 2.1.

Table 2.1: The Zeta converter parameters (see Fig. 2.1)

	Nominal Value
Output voltage, Vo	20 V
Duty ratio, D	0.5
Input DC source voltage, Vg	20 V
Input DC source internal resistance, rg	0.0 Ω
MOSFET Drain-Source resistance, rds	10 mΩ
Capacitor, C_1	100 μF
Capacitor C_1 Equivaluent Series Resistance (ESR), rC	0.19 Ω
Capacitor, C_2	220 μF
Capacitor C_2 Equivaluent Series Resistance (ESR), rC	0.095 Ω
Inductor, L_1	100 μH
Inductor ESR, rL_1	1 mΩ
Inductor, L_2	55 μH
Inductor ESR, rL_2	0.55 mΩ
Diode voltage drop, vD	0.7 V
Diode forward resistance, rD	10 mΩ
Load resistor, R	6 Ω
Switching Frequency, Fsw	100 KHz

The following MATLAB®code, extracts the steady-state operating point (average values of inductor currents and capacitor voltages) of the converter.

```
%This program calculates the DC operating point of a
%Zeta converter
clc
clear all

VG=20;        %Value of input DC source
rg=0;         %Internal resistance of input DC source
rds=.01;      %MOSFET on resistance
C1=100e-6;    %Capacitor C1 value
C2=220e-6;    %Capacitor C2 value
rC1=.19;      %Capacitor C1 Equivalent Series Resistance(ESR)
rC2=.095;     %Capacitor C2 Equivalent Series Resistance(ESR)
L1=100e-6;    %Inductor L1 value
L2=55e-6;     %Inductor L2 value
rL1=1e-3;     %Inductor L1 Equivalent Series Resistance(ESR)
rL2=.55e-3;   %Inductor L2 Equivalent Series Resistance(ESR)
rD=.01;       %Diode series resistance
VD=.7;        %Diode voltage drop
R=6;          %Load resistance
D=.5;         %Duty cylcle
IO=0;         %Average value of output current source
fsw=100e3;    %Switching frequency

syms iL1 iL2 vC1 vC2 io vg vD d
% iL1: Inductor L1 current
% iL2: Inductor L2 current
% vC1: Capacitor C1 voltage
% vC2: Capacitor C2 voltage
% io : Output current source
% vg : Input DC source
% vD : Diode voltage drop
% d  : Duty cycle

%Closed MOSFET Equations
diL1_dt_MOSFET_close=(-(rL1+rg+rds)*iL1-(rg+rds)*iL2+vg)/L1;
diL2_dt_MOSFET_close=(-(rg+rds)*iL1-(rg+rds+rC1+rL2+R*rC2/
```

```
   (R+rC2))*iL2+vC1-R/(R+rC2)*vC2+R*rC2/(R+rC2)*io+vg)/L2;
dvC1_dt_MOSFET_close=(-iL2)/C1;
dvC2_dt_MOSFET_close=(R/(R+rC2)*iL2-1/(R+rC2)*vC2-R/
   (R+rC2)*io)/C2;
vo_MOSFET_close=R*rC2/(R+rC2)*iL2+R/(R+rC2)*vC2-R*rC2/(R+rC2)*io;

%Opened MOSFET Equations
diL1_dt_MOSFET_open=(-(rL1+rC1+rD)*iL1-rD*iL2-vC1-vD)/L1;
diL2_dt_MOSFET_open=(-rD*iL1-(rD+rL2+R*rC2/(R+rC2))*iL2-R/
   (R+rC2)*vC2+R*rC2/(R+rC2)*io-vD)/L2;
dvC1_dt_MOSFET_open=(iL1)/C1;
dvC2_dt_MOSFET_open=(R/(R+rC2)*iL2-1/(R+rC2)*vC2-R/
   (R+rC2)*io)/C2;
vo_MOSFET_open=R*rC2/(R+rC2)*iL2+R/(R+rC2)*vC2-R*rC2/(R+rC2)*io;

%Averaging
averaged_diL1_dt=simplify(d*diL1_dt_MOSFET_close+(1-d)*
   diL1_dt_MOSFET_open);
averaged_diL2_dt=simplify(d*diL2_dt_MOSFET_close+(1-d)*
   diL2_dt_MOSFET_open);
averaged_dvC1_dt=simplify(d*dvC1_dt_MOSFET_close+(1-d)*
   dvC1_dt_MOSFET_open);
averaged_dvC2_dt=simplify(d*dvC2_dt_MOSFET_close+(1-d)*
   dvC2_dt_MOSFET_open);
averaged_vo=simplify(d*vo_MOSFET_close+(1-d)*vo_MOSFET_open);

%Substituting the steady values of input DC voltage source,
%Diode voltage drop, Duty cycle and output current source
%and calculating the DC operating point
right_side_of_averaged_diL1_dt=subs(averaged_diL1_dt,
   [vg vD d io],[VG VD D IO]);
right_side_of_averaged_diL2_dt=subs(averaged_diL2_dt,
   [vg vD d io],[VG VD D IO]);
right_side_of_averaged_dvC1_dt=subs(averaged_dvC1_dt,
   [vg vD d io],[VG VD D IO]);
right_side_of_averaged_dvC2_dt=subs(averaged_dvC2_dt,
   [vg vD d io],[VG VD D IO]);

DC_OPERATING_POINT=
```

```
solve(right_side_of_averaged_diL1_dt==0,
    right_side_of_averaged_diL2_dt==0,
    right_side_of_averaged_dvC1_dt==0,
    right_side_of_averaged_dvC2_dt==0,'iL1','iL2','vC1','vC2');

IL1=eval(DC_OPERATING_POINT.iL1);
IL2=eval(DC_OPERATING_POINT.iL2);
VC1=eval(DC_OPERATING_POINT.vC1);
VC2=eval(DC_OPERATING_POINT.vC2);
VO=eval(subs(averaged_vo,[iL1 iL2 vC1 vC2 io],
    [IL1 IL2 VC1 VC2 IO]));

disp('Operating point of converter')
disp('---------------------------')
disp('IL1(A)=')
disp(IL1)
disp('IL2(A)=')
disp(IL2)
disp('VC1(V)=')
disp(VC1)
disp('VC2(V)=')
disp(VC2)
disp('VO(V)=')
disp(VO)
disp('---------------------------')
```

After running the code, the result shown in Fig. 2.4 is obtained.

According to the analysis results, average currents through inductors L_1 and L_2 are 0.2593 A and 0.8683 A, respectively. The average voltage of capacitors C_1 and C_2 are, 5.2098 V and 5.2096 V, respectively. Obtained results can be checked with the aid of PLECS®. The schematic shown in Fig. 2.5 can be used for this purpose.

The PWM signal is produced with the aid of a "Sawtooth PWM" block (Fig. 2.6). Figure 2.7 shows the settings of the "Sawtooth PWM" block used in Fig. 2.5.

Average value of signals are calculated with the aid of "Discrete Mean Value" block (Fig. 2.8). Figure 2.9 shows the settings of the blocks used in Fig. 2.5.

The components values are defined in the "Initialization" tab of "Simulation Parameters" window. The "Simulation Parameters" can be reached with the aid of "Simulation" menu.

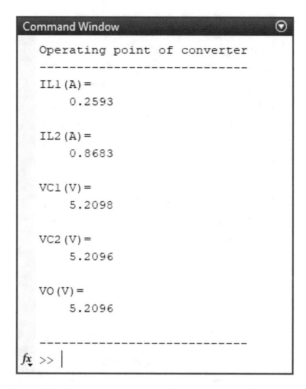

Figure 2.4: Average value of inductors currents and capacitors voltages.

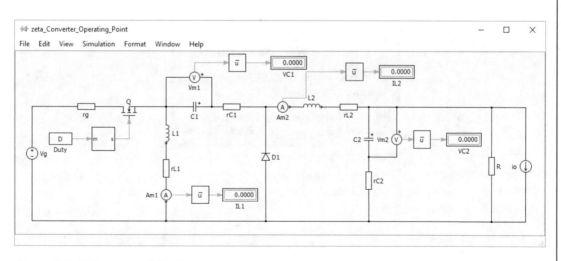

Figure 2.5: Schematic of the Zeta converter.

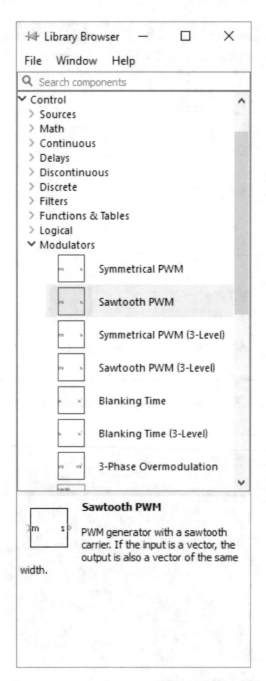

Figure 2.6: "Sawtooth PWM" block can be found under the "Modulators."

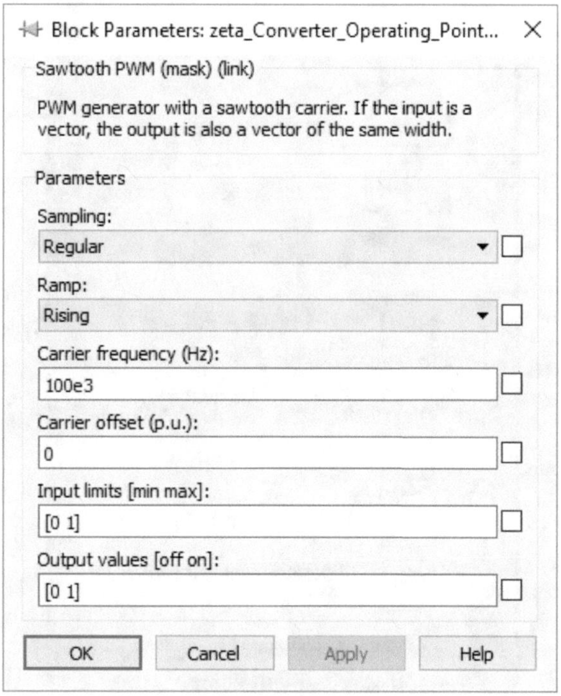

Figure 2.7: Setting of the "Sawtooth PWM" block used in Fig. 2.5.

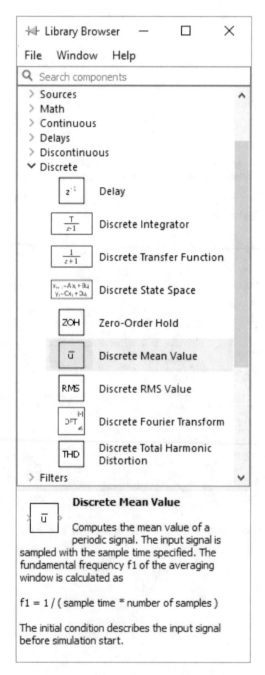

Figure 2.8: "Discrete Mean Value" block can be found under the "Discrete."

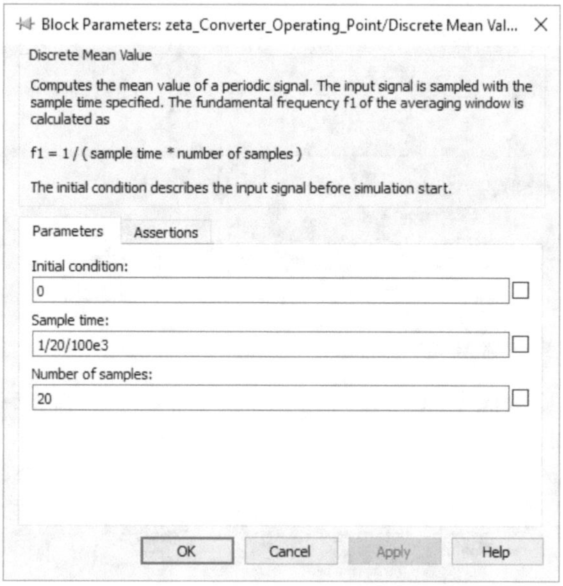

Figure 2.9: Setting of the "Discrete Mean Value" block used in Fig. 2.5.

After the components values are defined in the "Initialization" tab of "Simulation parameters" window, you can use them in the components values boxes. For instance, the "Resistance" box of the resistor connected to the input DC source, is filled with the variable named "rg." Figure 2.13 shows the simulation settings. The simulation can be run by clicking the "Start" or pressing the Ctrl+T. After running the simulation, the result shown in Fig. 2.15 will obtain. The obtained results are the same as the one obtained with the aid of MATLAB®code.

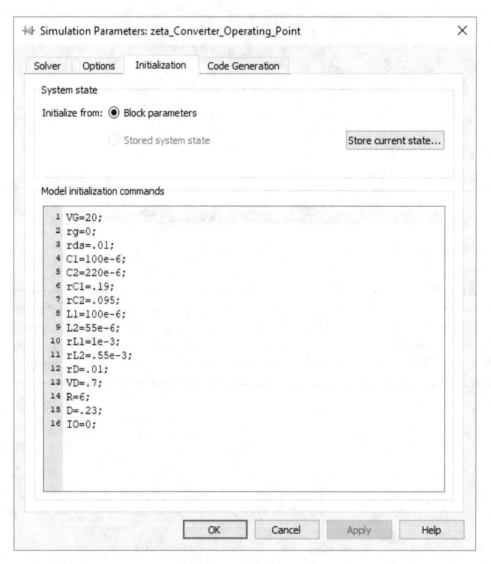

Figure 2.10: Converter components values.

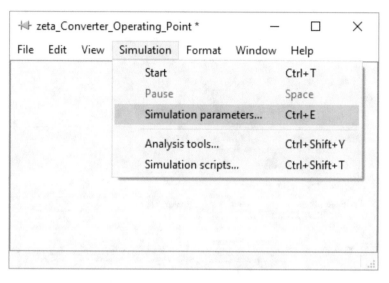

Figure 2.11: "Simulation Parameters" window will appear after clicking the "Simulation parameters... ."

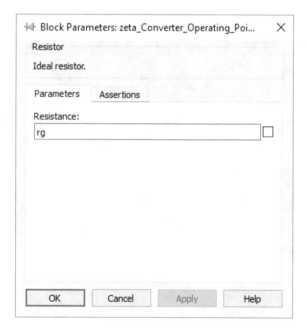

Figure 2.12: The resistance box is filled with a variable name instead of a numeric data. According to Fig. 2.10, the rg=0.

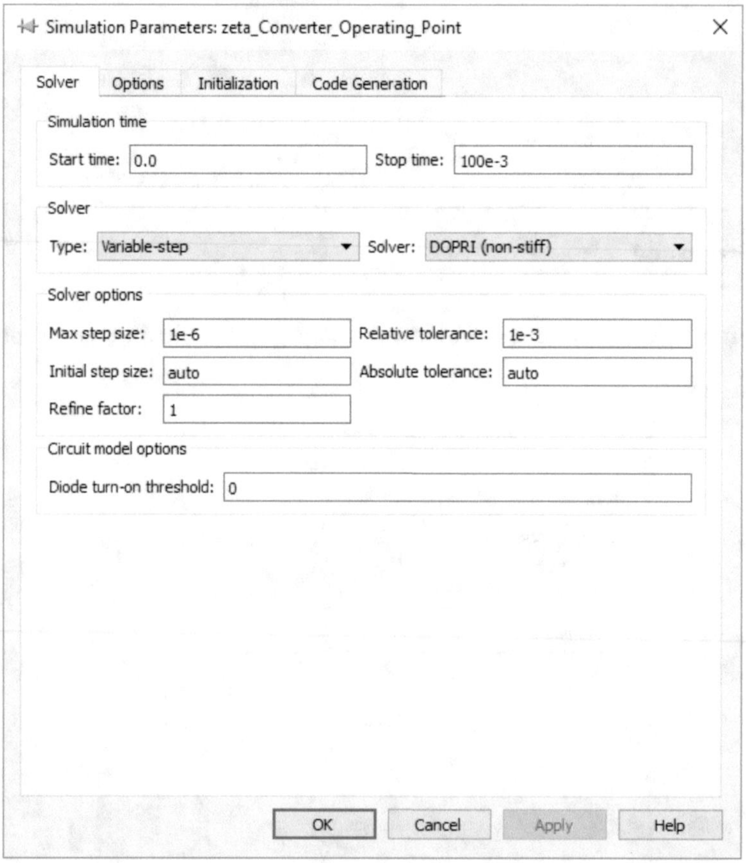

Figure 2.13: Simulation settings.

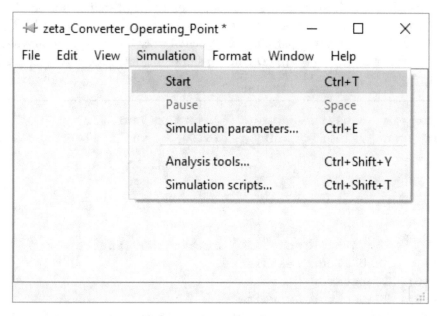

Figure 2.14: Click the "Start" to run the simulation.

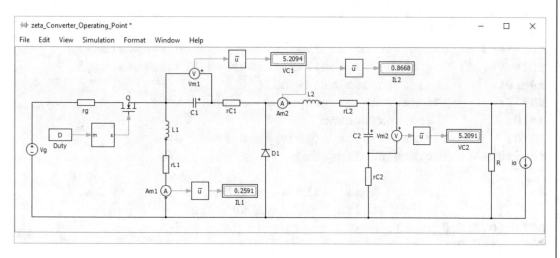

Figure 2.15: Obtained results. $I_{L_1} = 0.2591$ A, $I_{L_2} = 0.8668$ A, $V_{C_1} = 5.2094$ V, $V_{C_2} = 5.2091$ V.

2.4 DRAWING THE VOLTAGE GAIN RATIO

The following MATLAB®code compares the voltage gain ratio of studied non-ideal Zeta converter (i.e., a converter with parasitic components such as capacitors/inductors ESR) with the ideal (i.e., a converter without any parasitic components) case. The average output voltage of an ideal Zeta converter is $V_O = \frac{D}{1-D}V_{IN} \cdot D$ and V_{IN} shows the duty ratio and input DC voltage, respectively.

```
%This program shows the gain of converter(Vout/Vin)
%for different values of duty ratio(D) in the following range
%0.4..0.8
clc
clear all

VG=20;         %Value of input DC source
rg=0;          %Internal resistance of input DC source
rds=.01;       %MOSFET on resistance
C1=100e-6;     %Capacitor C1 value
C2=220e-6;     %Capacitor C2 value
rC1=.19;       %Capacitor C1 Equivalent Series Resistance(ESR)
rC2=.095;      %Capacitor C2 Equivalent Series Resistance(ESR)
L1=100e-6;     %Inductor L1 value
L2=55e-6;      %Inductor L2 value
rL1=1e-3;      %Inductor L1 Equivalent Series Resistance(ESR)
rL2=.55e-3;    %Inductor L2 Equivalent Series Resistance(ESR)
rD=.01;        %Diode series resistance
VD=.7;         %Diode voltage drop
R=10;          %Load resistance
IO=0;          %Average value of output current source
fsw=100e3;     %Switching frequency

M=[0];         %Voltage gain ratio is saved in this variable.
               %We initialize M with value of 0.
DUTY=[0.4:0.01:0.8];
N=length(DUTY);
n=0;

for D=DUTY
n=n+1;
disp('percentage of work done:')
```

```
disp(n/N*100)
disp('-----------------------')

syms iL1 iL2 vC1 vC2 io vg vD d
% iL1: Inductor L1 current
% iL2: Inductor L2 current
% vC1: Capacitor C1 voltage
% vC2: Capacitor C2 voltage
% io : Output current source
% vg : Input DC source
% vD : Diode voltage drop
% d : Duty cycle

%Closed MOSFET Equations
diL1_dt_MOSFET_close=(-(rL1+rg+rds)*iL1-(rg+rds)*iL2+vg)/L1;
diL2_dt_MOSFET_close=(-(rg+rds)*iL1-(rg+rds+rC1+rL2+R*rC2/
    (R+rC2))*iL2+vC1-R/(R+rC2)*vC2+R*rC2/(R+rC2)*io+vg)/L2;
dvC1_dt_MOSFET_close=(-iL2)/C1;
dvC2_dt_MOSFET_close=(R/(R+rC2)*iL2-1/(R+rC2)*vC2-R/
    (R+rC2)*io)/C2;
vo_MOSFET_close=R*rC2/(R+rC2)*iL2+R/(R+rC2)*vC2-R*rC2/(R+rC2)*io;

%Opened MOSFET Equations
diL1_dt_MOSFET_open=(-(rL1+rC1+rD)*iL1-rD*iL2-vC1-vD)/L1;
diL2_dt_MOSFET_open=(-rD*iL1-(rD+rL2+R*rC2/(R+rC2))*
    iL2-R/(R+rC2)*vC2+R*rC2/(R+rC2)*io-vD)/L2;
dvC1_dt_MOSFET_open=(iL1)/C1;
dvC2_dt_MOSFET_open=(R/(R+rC2)*iL2-1/(R+rC2)*
    vC2-R/(R+rC2)*io)/C2;
vo_MOSFET_open=R*rC2/(R+rC2)*iL2+R/(R+rC2)*vC2-R*rC2/(R+rC2)*io;

%Averaging
averaged_diL1_dt=simplify(d*diL1_dt_MOSFET_close+(1-d)*
    diL1_dt_MOSFET_open);
averaged_diL2_dt=simplify(d*diL2_dt_MOSFET_close+(1-d)*
    diL2_dt_MOSFET_open);
averaged_dvC1_dt=simplify(d*dvC1_dt_MOSFET_close+(1-d)*
    dvC1_dt_MOSFET_open);
averaged_dvC2_dt=simplify(d*dvC2_dt_MOSFET_close+(1-d)*
```

```matlab
    dvC2_dt_MOSFET_open);
averaged_vo=simplify(d*vo_MOSFET_close+(1-d)*vo_MOSFET_open);

%Substituting the steady values of input DC voltage source,
%Diode voltage drop, Duty cycle and output current source
%and calculating the DC operating point
right_side_of_averaged_diL1_dt=subs(averaged_diL1_dt,
    [vg vD d io],[VG VD D IO]);
right_side_of_averaged_diL2_dt=subs(averaged_diL2_dt,
    [vg vD d io],[VG VD D IO]);
right_side_of_averaged_dvC1_dt=subs(averaged_dvC1_dt,
    [vg vD d io],[VG VD D IO]);
right_side_of_averaged_dvC2_dt=subs(averaged_dvC2_dt,
    [vg vD d io],[VG VD D IO]);

DC_OPERATING_POINT=
solve(right_side_of_averaged_diL1_dt==0,
    right_side_of_averaged_diL2_dt==0,
    right_side_of_averaged_dvC1_dt==0,
    right_side_of_averaged_dvC2_dt==0,'iL1','iL2','vC1','vC2');

IL1=eval(DC_OPERATING_POINT.iL1);
IL2=eval(DC_OPERATING_POINT.iL2);
VC1=eval(DC_OPERATING_POINT.vC1);
VC2=eval(DC_OPERATING_POINT.vC2);
VO=eval(subs(averaged_vo,[iL1 iL2 vC1 vC2 io],
    [IL1 IL2 VC1 VC2 IO]));

M=[M;VO/VG]; %Voltage Gain Ratio (Vout/Vin)
end

M(1)=[];      %Clear the initialization
plot(DUTY,M),grid minor,
title('Voltage gain ratio of Zeta converter')
xlabel('Duty ratio')
ylabel('Voltage gain ratio')
hold on
plot(DUTY,DUTY./(1-DUTY),'r')
legend('Non-Ideal','Ideal')
```

After running the code, the result shown in Fig. 2.16 will be obtained. As expected, the voltage gain of non-ideal converter is lower than the ideal case.

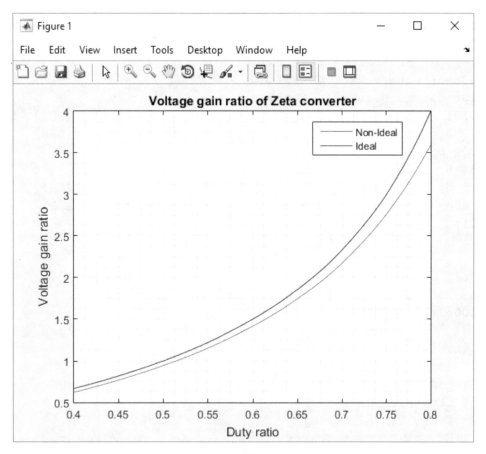

Figure 2.16: Comparison of voltage gain ratio ($\frac{V_{OUT}}{V_{IN}}$) for ideal and non-ideal Zeta converters.

2.5 OBTAINING THE SMALL SIGNAL TRANSFER FUNCTIONS OF CONVERTER

The following code extracts the small signal transfer functions of the studied Zeta converter. We assume that the components have no variations. Effects of components variations are studied in the next sections.

```
%This program calculates the small signal transfer
%functions of Zeta converter
```

```
clc
clear all

VG=20;        %Value of input DC source
rg=0;         %Internal resistance of input DC source
rds=.01;      %MOSFET on resistance
C1=100e-6;    %Capacitor C1 value
C2=220e-6;    %Capacitor C2 value
rC1=.19;      %Capacitor C1 Equivalent Series Resistance(ESR)
rC2=.095;     %Capacitor C2 Equivalent Series Resistance(ESR)
L1=100e-6;    %Inductor L1 value
L2=55e-6;     %Inductor L2 value
rL1=1e-3;     %Inductor L1 Equivalent Series Resistance(ESR)
rL2=.55e-3;   %Inductor L2 Equivalent Series Resistance(ESR)
rD=.01;       %Diode series resistance
VD=.7;        %Diode voltage drop
R=6;          %Load resistance
D=.23;        %Duty cylcle
IO=0;         %Average value of output current source
fsw=100e3;    %Switching frequency

syms iL1 iL2 vC1 vC2 io vg vD d
% iL1: Inductor L1 current
% iL2: Inductor L2 current
% vC1: Capacitor C1 voltage
% vC2: Capacitor C2 voltage
% io : Output current source
% vg : Input DC source
% vD : Diode voltage drop
% d  : Duty cycle

%Closed MOSFET Equations
diL1_dt_MOSFET_close=(-(rL1+rg+rds)*iL1-(rg+rds)*iL2+vg)/L1;
diL2_dt_MOSFET_close=(-(rg+rds)*iL1-(rg+rds+rC1+rL2+R*rC2/
    (R+rC2))*iL2+vC1-R/(R+rC2)*vC2+R*rC2/(R+rC2)*io+vg)/L2;
dvC1_dt_MOSFET_close=(-iL2)/C1;
dvC2_dt_MOSFET_close=(R/(R+rC2)*iL2-1/(R+rC2)*vC2-R/(R+rC2)*
    io)/C2;
vo_MOSFET_close=R*rC2/(R+rC2)*iL2+R/(R+rC2)*vC2-R*rC2/(R+rC2)*io;
```

```
%Opened MOSFET Equations
diL1_dt_MOSFET_open=(-(rL1+rC1+rD)*iL1-rD*iL2-vC1-vD)/L1;
diL2_dt_MOSFET_open=(-rD*iL1-(rD+rL2+R*rC2/(R+rC2))*iL2-R/
    (R+rC2)*vC2+R*rC2/(R+rC2)*io-vD)/L2;
dvC1_dt_MOSFET_open=(iL1)/C1;
dvC2_dt_MOSFET_open=(R/(R+rC2)*iL2-1/(R+rC2)*vC2-R/
    (R+rC2)*io)/C2;
vo_MOSFET_open=R*rC2/(R+rC2)*iL2+R/(R+rC2)*vC2-R*rC2/(R+rC2)*io;

%Averaging
averaged_diL1_dt=simplify(d*diL1_dt_MOSFET_close+(1-d)*
    diL1_dt_MOSFET_open);
averaged_diL2_dt=simplify(d*diL2_dt_MOSFET_close+(1-d)*
    diL2_dt_MOSFET_open);
averaged_dvC1_dt=simplify(d*dvC1_dt_MOSFET_close+(1-d)*
    dvC1_dt_MOSFET_open);
averaged_dvC2_dt=simplify(d*dvC2_dt_MOSFET_close+(1-d)*
    dvC2_dt_MOSFET_open);
averaged_vo=simplify(d*vo_MOSFET_close+(1-d)*vo_MOSFET_open);

%Substituting the steady values of input DC voltage source,
%Diode voltage drop, Duty cycle and output current source
%and calculating the DC operating point
right_side_of_averaged_diL1_dt=subs(averaged_diL1_dt,
    [vg vD d io],[VG VD D IO]);
right_side_of_averaged_diL2_dt=subs(averaged_diL2_dt,
    [vg vD d io],[VG VD D IO]);
right_side_of_averaged_dvC1_dt=subs(averaged_dvC1_dt,
    [vg vD d io],[VG VD D IO]);
right_side_of_averaged_dvC2_dt=subs(averaged_dvC2_dt,
    [vg vD d io],[VG VD D IO]);

DC_OPERATING_POINT=
solve(right_side_of_averaged_diL1_dt==0,
    right_side_of_averaged_diL2_dt==0,
    right_side_of_averaged_dvC1_dt==0,
    right_side_of_averaged_dvC2_dt==0,'iL1','iL2','vC1','vC2');
```

```
IL1=eval(DC_OPERATING_POINT.iL1);
IL2=eval(DC_OPERATING_POINT.iL2);
VC1=eval(DC_OPERATING_POINT.vC1);
VC2=eval(DC_OPERATING_POINT.vC2);
VO=eval(subs(averaged_vo,[iL1 iL2 vC1 vC2 io],
    [IL1 IL2 VC1 VC2 IO]));

disp('Operating point of converter')
disp('----------------------------')
disp('IL1(A)=')
disp(IL1)
disp('IL2(A)=')
disp(IL2)
disp('VC1(V)=')
disp(VC1)
disp('VC2(V)=')
disp(VC2)
disp('VO(V)=')
disp(VO)
disp('----------------------------')

%Linearizing the averaged equations around the DC
%operating point. We want to obtain the matrix
%A, B, C, and D
%        .
%        x=Ax+Bu
%        y=Cx+Du
%
%where,
%        x=[iL1 iL2 vC1 vC2]'
%        u=[io vg d]'
%Since we used the variables D for steady state duty
%ratio and C to show the capacitors values we use
%AA, BB, CC, and DD instead of A, B, C, and D.

% Calculating the matrix A
A11=subs(simplify(diff(averaged_diL1_dt,iL1)),
    [iL1 iL2 vC1 vC2 d io],[IL1 IL2 VC1 VC2 D IO]);
A12=subs(simplify(diff(averaged_diL1_dt,iL2)),
```

```
    [iL1 iL2 vC1 vC2 d io],[IL1 IL2 VC1 VC2 D IO]);
A13=subs(simplify(diff(averaged_diL1_dt,vC1)),
    [iL1 iL2 vC1 vC2 d io],[IL1 IL2 VC1 VC2 D IO]);
A14=subs(simplify(diff(averaged_diL1_dt,vC2)),
    [iL1 iL2 vC1 vC2 d io],[IL1 IL2 VC1 VC2 D IO]);

A21=subs(simplify(diff(averaged_diL2_dt,iL1)),
    [iL1 iL2 vC1 vC2 d io],[IL1 IL2 VC1 VC2 D IO]);
A22=subs(simplify(diff(averaged_diL2_dt,iL2)),
    [iL1 iL2 vC1 vC2 d io],[IL1 IL2 VC1 VC2 D IO]);
A23=subs(simplify(diff(averaged_diL2_dt,vC1)),
    [iL1 iL2 vC1 vC2 d io],[IL1 IL2 VC1 VC2 D IO]);
A24=subs(simplify(diff(averaged_diL2_dt,vC2)),
    [iL1 iL2 vC1 vC2 d io],[IL1 IL2 VC1 VC2 D IO]);

A31=subs(simplify(diff(averaged_dvC1_dt,iL1)),
    [iL1 iL2 vC1 vC2 d io],[IL1 IL2 VC1 VC2 D IO]);
A32=subs(simplify(diff(averaged_dvC1_dt,iL2)),
    [iL1 iL2 vC1 vC2 d io],[IL1 IL2 VC1 VC2 D IO]);
A33=subs(simplify(diff(averaged_dvC1_dt,vC1)),
    [iL1 iL2 vC1 vC2 d io],[IL1 IL2 VC1 VC2 D IO]);
A34=subs(simplify(diff(averaged_dvC1_dt,vC2)),
    [iL1 iL2 vC1 vC2 d io],[IL1 IL2 VC1 VC2 D IO]);

A41=subs(simplify(diff(averaged_dvC2_dt,iL1)),
    [iL1 iL2 vC1 vC2 d io],[IL1 IL2 VC1 VC2 D IO]);
A42=subs(simplify(diff(averaged_dvC2_dt,iL2)),
    [iL1 iL2 vC1 vC2 d io],[IL1 IL2 VC1 VC2 D IO]);
A43=subs(simplify(diff(averaged_dvC2_dt,vC1)),
    [iL1 iL2 vC1 vC2 d io],[IL1 IL2 VC1 VC2 D IO]);
A44=subs(simplify(diff(averaged_dvC2_dt,vC2)),
    [iL1 iL2 vC1 vC2 d io],[IL1 IL2 VC1 VC2 D IO]);

AA=eval([A11 A12 A13 A14;
         A21 A22 A23 A24;
         A31 A32 A33 A34;
         A41 A42 A43 A44]);

%Calculating the matrix B
```

```
B11=subs(simplify(diff(averaged_diL1_dt,io)),
    [iL1 iL2 vC1 vC2 d vD io vg],[IL1 IL2 VC1 VC2 D VD IO VG]);
B12=subs(simplify(diff(averaged_diL1_dt,vg)),
    [iL1 iL2 vC1 vC2 d vD io vg],[IL1 IL2 VC1 VC2 D VD IO VG]);
B13=subs(simplify(diff(averaged_diL1_dt,d)),
    [iL1 iL2 vC1 vC2 d vD io vg],[IL1 IL2 VC1 VC2 D VD IO VG]);

B21=subs(simplify(diff(averaged_diL2_dt,io)),
    [iL1 iL2 vC1 vC2 d vD io vg],[IL1 IL2 VC1 VC2 D VD IO VG]);
B22=subs(simplify(diff(averaged_diL2_dt,vg)),
    [iL1 iL2 vC1 vC2 d vD io vg],[IL1 IL2 VC1 VC2 D VD IO VG]);
B23=subs(simplify(diff(averaged_diL2_dt,d)),
    [iL1 iL2 vC1 vC2 d vD io vg],[IL1 IL2 VC1 VC2 D VD IO VG]);

B31=subs(simplify(diff(averaged_dvC1_dt,io)),
    [iL1 iL2 vC1 vC2 d vD io vg],[IL1 IL2 VC1 VC2 D VD IO VG]);
B32=subs(simplify(diff(averaged_dvC1_dt,vg)),
    [iL1 iL2 vC1 vC2 d vD io vg],[IL1 IL2 VC1 VC2 D VD IO VG]);
B33=subs(simplify(diff(averaged_dvC1_dt,d)),
    [iL1 iL2 vC1 vC2 d vD io vg],[IL1 IL2 VC1 VC2 D VD IO VG]);

B41=subs(simplify(diff(averaged_dvC2_dt,io)),
    [iL1 iL2 vC1 vC2 d vD io vg],[IL1 IL2 VC1 VC2 D VD IO VG]);
B42=subs(simplify(diff(averaged_dvC2_dt,vg)),
    [iL1 iL2 vC1 vC2 d vD io vg],[IL1 IL2 VC1 VC2 D VD IO VG]);
B43=subs(simplify(diff(averaged_dvC2_dt,d)),
    [iL1 iL2 vC1 vC2 d vD io vg],[IL1 IL2 VC1 VC2 D VD IO VG]);

BB=eval([B11 B12 B13;
         B21 B22 B23;
         B31 B32 B33;
         B41 B42 B43]);

%Calculating the matrix C
C11=subs(simplify(diff(averaged_vo,iL1)),[iL1 iL2 vC1 vC2 d io],
    [IL1 IL2 VC1 VC2 D IO]);
C12=subs(simplify(diff(averaged_vo,iL2)),[iL1 iL2 vC1 vC2 d io],
    [IL1 IL2 VC1 VC2 D IO]);
C13=subs(simplify(diff(averaged_vo,vC1)),[iL1 iL2 vC1 vC2 d io],
```

```
   [IL1 IL2 VC1 VC2 D IO]);
C14=subs(simplify(diff(averaged_vo,vC2)),[iL1 iL2 vC1 vC2 d io],
   [IL1 IL2 VC1 VC2 D IO]);

CC=eval([C11 C12 C13 C14]);

D11=subs(simplify(diff(averaged_vo,io)),
   [iL1 iL2 vC1 vC2 d vD io vg],[IL1 IL2 VC1 VC2 D VD IO VG]);
D12=subs(simplify(diff(averaged_vo,vg)),
   [iL1 iL2 vC1 vC2 d vD io vg],[IL1 IL2 VC1 VC2 D VD IO VG]);
D13=subs(simplify(diff(averaged_vo,d)),
   [iL1 iL2 vC1 vC2 d vD io vg],[IL1 IL2 VC1 VC2 D VD IO VG]);

%Calculating the matrix D
DD=eval([D11 D12 D13]);

%Producing the State Space Model and obtaining the small
%signal transfer functions
sys=ss(AA,BB,CC,DD);
sys.inputname={'io';'vg';'d'};
sys.outputname={'vo'};

vo_io=tf(sys(1,1)); %Output impedance transfer function
                    %vo(s)/io(s)
vo_vg=tf(sys(1,2)); %vo(s)/vg(s)
vo_d=tf(sys(1,3));  %Control-to-output(vo(s)/d(s))

%drawing the Bode diagrams
figure(1)
bode(vo_io),grid minor,title('vo(s)/io(s)')

figure(2)
bode(vo_vg),grid minor,title('vo(s)/vg(s)')

figure(3)
bode(vo_d),grid minor,title('vo(s)/d(s)')
```

After running the code, the results shown in Figs. 2.17, 2.18, and 2.19 are obtained.

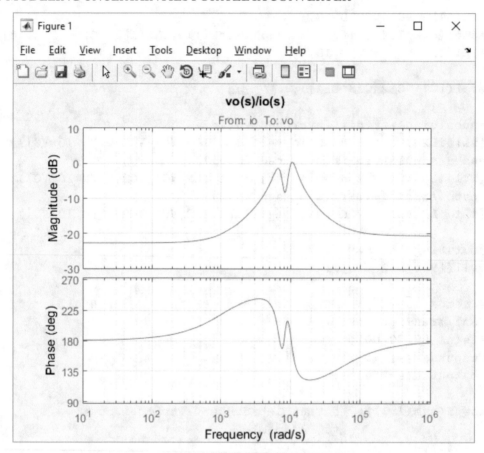

Figure 2.17: Bode diagram of the $\frac{v_o(s)}{i_o(s)}$ transfer function.

Algebraic form of transfer functions can be seen in MATLAB®command prompt with the aid of commands shown in Fig. 2.20.

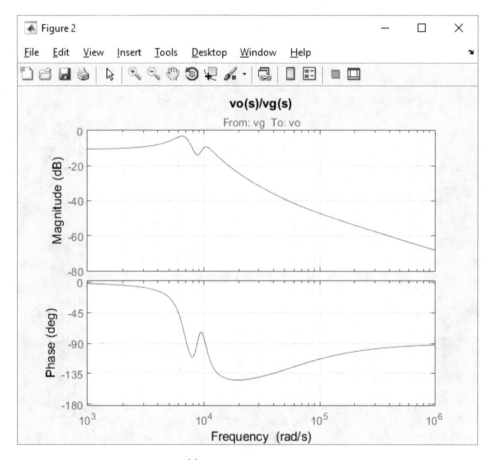

Figure 2.18: Bode diagram of the $\frac{v_o(s)}{v_g(s)}$ transfer function.

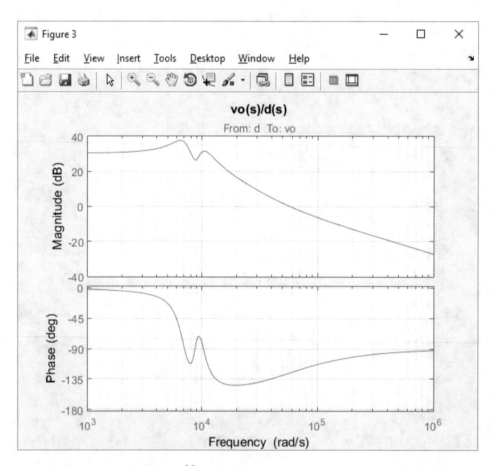

Figure 2.19: Bode diagram of the $\frac{v_o(s)}{d(s)}$ transfer function.

```
Command Window                                                    ⊙

  >> zpk(vo_io)

  ans =

    From input "io" to output "vo":
    -0.093519 (s+4.785e04) (s+1163) (s^2 + 1396s + 6.882e07)
    --------------------------------------------------------
         (s^2 + 2239s + 4.76e07) (s^2 + 2767s + 1.026e08)

  Continuous-time zero/pole/gain model.

  >> zpk(vo_vg)

  ans =

    From input "vg" to output "vo":
        391.08 (s+4.785e04) (s^2 + 1473s + 7.7e07)
    ------------------------------------------------
    (s^2 + 2239s + 4.76e07) (s^2 + 2767s + 1.026e08)

  Continuous-time zero/pole/gain model.

  >> zpk(vo_d)

  ans =

    From input "d" to output "vo":
      43775 (s+4.785e04) (s^2 + 1371s + 7.696e07)
    ------------------------------------------------
    (s^2 + 2239s + 4.76e07) (s^2 + 2767s + 1.026e08)

  Continuous-time zero/pole/gain model.

fx >> |
```

Figure 2.20: Obtained transfer functions equations.

2.6   EFFECT OF LOAD CHANGES ON THE SMALL SIGNAL TRANSFER FUNCTIONS

We study the effect of output load variations on the small signal transfer functions of the studied converter. It is assumed that output load changes between 1 Ω and 11 Ω. Table 2.2 shows the components values used in this study.

Table 2.2: The Zeta converter parameters (see Fig. 2.1)

	Nominal Value
Output voltage, Vo	20 V
Duty ratio, D	0.5
Input DC source voltage, Vg	20 V
Input DC source internal resistance, rg	0.0 Ω
MOSFET Drain-Source resistance, rds	10 mΩ
Capacitor, C_1	100 μF
Capacitor C_1 Equivaluent Series Resistance (ESR), rC	0.19 Ω
Capacitor, C_2	220 μF
Capacitor C_2 Equivaluent Series Resistance (ESR), rC	0.095 Ω
Inductor, L_1	100 μH
Inductor ESR, rL_1	1 mΩ
Inductor, L_2	55 μH
Inductor ESR, rL_2	0.55 mΩ
Diode voltage drop, vD	0.7 V
Diode forward resistance, rD	10 mΩ
Load resistor, R	**1 Ω<R <11 Ω**
Switching Frequency, Fsw	100 KHz

The following MATLAB®code uses a loop to extracts the small signal transfer functions for different values of loads. The heart of this code is the same as the previous one. Only a loop is added to supply the code with different values of load. One can use more nested loops in order to study the effect of other components changes.

```
%This program study the effect of change in output load
%Rnominal=6 ohm
%1 ohm < R < 11 ohm
clc
```

```
clear all
close all

%Desired load range for sweep
Rnominal=6;
Rmin=1;
R_delta=.5;
Rmax=11;

VG=20;        %Value of input DC source
rg=0;         %Internal resistance of input DC source
rds=.01;      %MOSFET on resistance
C1=100e-6;    %Capacitor C1 value
C2=220e-6;    %Capacitor C2 value
rC1=.19;      %Capacitor C1 Equivalent Series Resistance(ESR)
rC2=.095;     %Capacitor C2 Equivalent Series Resistance(ESR)
L1=100e-6;    %Inductor L1 value
L2=55e-6;     %Inductor L2 value
rL1=1e-3;     %Inductor L1 Equivalent Series Resistance(ESR)
rL2=.55e-3;   %Inductor L2 Equivalent Series Resistance(ESR)
rD=.01;       %Diode series resistance
VD=.7;        %Diode voltage drop
D=.23;        %Duty cylcle
IO=0;         %Average value of output current source
fsw=100e3;    %Switching frequency

n=0;
N=length({[]}Rmin:R_delta:Rmax{[]});

for R={[]}Rmin:R_delta:Rmax{[]}

n=n+1;
disp('Percentage of work done:')
disp(n/N*100) %shows the progress of the loop

syms iL1 iL2 vC1 vC2 io vg vD d
% iL1: Inductor L1 current
% iL2: Inductor L2 current
% vC1: Capacitor C1 voltage
```

```
% vC2: Capacitor C2 voltage
% io : Output current source
% vg : Input DC source
% vD : Diode voltage drop
% d : Duty cycle

%Closed MOSFET Equations
diL1_dt_MOSFET_close=(-(rL1+rg+rds)*iL1-(rg+rds)*iL2+vg)/L1;
diL2_dt_MOSFET_close=(-(rg+rds)*iL1-(rg+rds+rC1+rL2+R*rC2/
    (R+rC2))*iL2+vC1-R/(R+rC2)*vC2+R*rC2/(R+rC2)*io+vg)/L2;
dvC1_dt_MOSFET_close=(-iL2)/C1;
dvC2_dt_MOSFET_close=(R/(R+rC2)*iL2-1/(R+rC2)*vC2-R/
    (R+rC2)*io)/C2;
vo_MOSFET_close=R*rC2/(R+rC2)*iL2+R/(R+rC2)*vC2-R*rC2/(R+rC2)*io;

%Opened MOSFET Equations
diL1_dt_MOSFET_open=(-(rL1+rC1+rD)*iL1-rD*iL2-vC1-vD)/L1;
diL2_dt_MOSFET_open=(-rD*iL1-(rD+rL2+R*rC2/(R+rC2))*iL2-R/
    (R+rC2)*vC2+R*rC2/(R+rC2)*io-vD)/L2;
dvC1_dt_MOSFET_open=(iL1)/C1;
dvC2_dt_MOSFET_open=(R/(R+rC2)*iL2-1/(R+rC2)*vC2-R/
    (R+rC2)*io)/C2;
vo_MOSFET_open=R*rC2/(R+rC2)*iL2+R/(R+rC2)*vC2-R*rC2/(R+rC2)*io;

%Averaging
averaged_diL1_dt=simplify(d*diL1_dt_MOSFET_close+(1-d)*
    diL1_dt_MOSFET_open);
averaged_diL2_dt=simplify(d*diL2_dt_MOSFET_close+(1-d)*
    diL2_dt_MOSFET_open);
averaged_dvC1_dt=simplify(d*dvC1_dt_MOSFET_close+(1-d)*
    dvC1_dt_MOSFET_open);
averaged_dvC2_dt=simplify(d*dvC2_dt_MOSFET_close+(1-d)*
    dvC2_dt_MOSFET_open);
averaged_vo=simplify(d*vo_MOSFET_close+(1-d)*vo_MOSFET_open);

%Substituting the steady values of input DC voltage source,
%Diode voltage drop, Duty cycle and output current source
%and calculating the DC operating point
right_side_of_averaged_diL1_dt=subs(averaged_diL1_dt,
```

```
    [vg vD d io],[VG VD D IO]);
right_side_of_averaged_diL2_dt=subs(averaged_diL2_dt,
    [vg vD d io],[VG VD D IO]);
right_side_of_averaged_dvC1_dt=subs(averaged_dvC1_dt,
    [vg vD d io],[VG VD D IO]);
right_side_of_averaged_dvC2_dt=subs(averaged_dvC2_dt,
    [vg vD d io],[VG VD D IO]);

DC_OPERATING_POINT=
solve(right_side_of_averaged_diL1_dt==0,
    right_side_of_averaged_diL2_dt==0,
    right_side_of_averaged_dvC1_dt==0,
    right_side_of_averaged_dvC2_dt==0,'iL1','iL2','vC1','vC2');

IL1=eval(DC_OPERATING_POINT.iL1);
IL2=eval(DC_OPERATING_POINT.iL2);
VC1=eval(DC_OPERATING_POINT.vC1);
VC2=eval(DC_OPERATING_POINT.vC2);
VO=eval(subs(averaged_vo,[iL1 iL2 vC1 vC2 io],
    [IL1 IL2 VC1 VC2 IO]));

%Linearizing the averaged equations around the DC
%operating point. We want to obtain the matrix A, B, C, and D
%           .
%       x=Ax+Bu
%       y=Cx+Du
%
%where,
%       x={[}iL1 iL2 vC1 vC2{]}'
%       u={[}io vg d{]}'
%Since we used the variables D for steady state duty
%ratio and C to show the capacitors values we use AA,
%BB, CC, and DD instead of A, B, C, and D.

% Calculating the matrix A
A11=subs(simplify(diff(averaged_diL1_dt,iL1)),
    [iL1 iL2 vC1 vC2 d io],[IL1 IL2 VC1 VC2 D IO]);
A12=subs(simplify(diff(averaged_diL1_dt,iL2)),
    [iL1 iL2 vC1 vC2 d io],[IL1 IL2 VC1 VC2 D IO]);
```

```
A13=subs(simplify(diff(averaged_diL1_dt,vC1)),
    [iL1 iL2 vC1 vC2 d io],[IL1 IL2 VC1 VC2 D IO]);
A14=subs(simplify(diff(averaged_diL1_dt,vC2)),
    [iL1 iL2 vC1 vC2 d io],[IL1 IL2 VC1 VC2 D IO]);

A21=subs(simplify(diff(averaged_diL2_dt,iL1)),
    [iL1 iL2 vC1 vC2 d io],[IL1 IL2 VC1 VC2 D IO]);
A22=subs(simplify(diff(averaged_diL2_dt,iL2)),
    [iL1 iL2 vC1 vC2 d io],[IL1 IL2 VC1 VC2 D IO]);
A23=subs(simplify(diff(averaged_diL2_dt,vC1)),
    [iL1 iL2 vC1 vC2 d io],[IL1 IL2 VC1 VC2 D IO]);
A24=subs(simplify(diff(averaged_diL2_dt,vC2)),
    [iL1 iL2 vC1 vC2 d io],[IL1 IL2 VC1 VC2 D IO]);

A31=subs(simplify(diff(averaged_dvC1_dt,iL1)),
    [iL1 iL2 vC1 vC2 d io],[IL1 IL2 VC1 VC2 D IO]);
A32=subs(simplify(diff(averaged_dvC1_dt,iL2)),
    [iL1 iL2 vC1 vC2 d io],[IL1 IL2 VC1 VC2 D IO]);
A33=subs(simplify(diff(averaged_dvC1_dt,vC1)),
    [iL1 iL2 vC1 vC2 d io],[IL1 IL2 VC1 VC2 D IO]);
A34=subs(simplify(diff(averaged_dvC1_dt,vC2)),
    [iL1 iL2 vC1 vC2 d io],[IL1 IL2 VC1 VC2 D IO]);

A41=subs(simplify(diff(averaged_dvC2_dt,iL1)),
    [iL1 iL2 vC1 vC2 d io],[IL1 IL2 VC1 VC2 D IO]);
A42=subs(simplify(diff(averaged_dvC2_dt,iL2)),
    [iL1 iL2 vC1 vC2 d io],[IL1 IL2 VC1 VC2 D IO]);
A43=subs(simplify(diff(averaged_dvC2_dt,vC1)),
    [iL1 iL2 vC1 vC2 d io],[IL1 IL2 VC1 VC2 D IO]);
A44=subs(simplify(diff(averaged_dvC2_dt,vC2)),
    [iL1 iL2 vC1 vC2 d io],[IL1 IL2 VC1 VC2 D IO]);

AA=eval([A11 A12 A13 A14;
         A21 A22 A23 A24;
         A31 A32 A33 A34;
         A41 A42 A43 A44]);

%Calculating the matrix B
B11=subs(simplify(diff(averaged_diL1_dt,io)),
```

```
        [iL1 iL2 vC1 vC2 d vD io vg],[IL1 IL2 VC1 VC2 D VD IO VG]);
B12=subs(simplify(diff(averaged_diL1_dt,vg)),
        [iL1 iL2 vC1 vC2 d vD io vg],[IL1 IL2 VC1 VC2 D VD IO VG]);
B13=subs(simplify(diff(averaged_diL1_dt,d)),
        [iL1 iL2 vC1 vC2 d vD io vg],[IL1 IL2 VC1 VC2 D VD IO VG]);

B21=subs(simplify(diff(averaged_diL2_dt,io)),
        [iL1 iL2 vC1 vC2 d vD io vg],[IL1 IL2 VC1 VC2 D VD IO VG]);
B22=subs(simplify(diff(averaged_diL2_dt,vg)),
        [iL1 iL2 vC1 vC2 d vD io vg],[IL1 IL2 VC1 VC2 D VD IO VG]);
B23=subs(simplify(diff(averaged_diL2_dt,d)),
        [iL1 iL2 vC1 vC2 d vD io vg],[IL1 IL2 VC1 VC2 D VD IO VG]);

B31=subs(simplify(diff(averaged_dvC1_dt,io)),
        [iL1 iL2 vC1 vC2 d vD io vg],[IL1 IL2 VC1 VC2 D VD IO VG]);
B32=subs(simplify(diff(averaged_dvC1_dt,vg)),
        [iL1 iL2 vC1 vC2 d vD io vg],[IL1 IL2 VC1 VC2 D VD IO VG]);
B33=subs(simplify(diff(averaged_dvC1_dt,d)),
        [iL1 iL2 vC1 vC2 d vD io vg],[IL1 IL2 VC1 VC2 D VD IO VG]);

B41=subs(simplify(diff(averaged_dvC2_dt,io)),
        [iL1 iL2 vC1 vC2 d vD io vg],[IL1 IL2 VC1 VC2 D VD IO VG]);
B42=subs(simplify(diff(averaged_dvC2_dt,vg)),
        [iL1 iL2 vC1 vC2 d vD io vg],[IL1 IL2 VC1 VC2 D VD IO VG]);
B43=subs(simplify(diff(averaged_dvC2_dt,d)),
        [iL1 iL2 vC1 vC2 d vD io vg],[IL1 IL2 VC1 VC2 D VD IO VG]);

BB=eval([B11 B12 B13;
         B21 B22 B23;
         B31 B32 B33;
         B41 B42 B43{]});

%Calculating the matrix C
C11=subs(simplify(diff(averaged_vo,iL1)),
        [iL1 iL2 vC1 vC2 d io],[IL1 IL2 VC1 VC2 D IO]);
C12=subs(simplify(diff(averaged_vo,iL2)),
        [iL1 iL2 vC1 vC2 d io],[IL1 IL2 VC1 VC2 D IO]);
C13=subs(simplify(diff(averaged_vo,vC1)),
        [iL1 iL2 vC1 vC2 d io],[IL1 IL2 VC1 VC2 D IO]);
```

```
C14=subs(simplify(diff(averaged_vo,vC2)),
    [iL1 iL2 vC1 vC2 d io],[IL1 IL2 VC1 VC2 D IO]);


CC=eval([C11 C12 C13 C14]);


D11=subs(simplify(diff(averaged_vo,io)),
    [iL1 iL2 vC1 vC2 d vD io vg],[IL1 IL2 VC1 VC2 D VD IO VG]);
D12=subs(simplify(diff(averaged_vo,vg)),
    [iL1 iL2 vC1 vC2 d vD io vg],[IL1 IL2 VC1 VC2 D VD IO VG]);
D13=subs(simplify(diff(averaged_vo,d)),
    [iL1 iL2 vC1 vC2 d vD io vg],[IL1 IL2 VC1 VC2 D VD IO VG]);


%Calculating the matrix D
DD=eval([D11 D12 D13]);


%Producing the State Space Model and obtaining the
%small signal transfer functions
sys=ss(AA,BB,CC,DD);
sys.inputname={'io';'vg';'d'};
sys.outputname={'vo'};


vo_io=tf(sys(1,1)); %Output impedance transfer function
                    %vo(s)/io(s)
vo_vg=tf(sys(1,2)); %vo(s)/vg(s)
vo_d=tf(sys(1,3));  %Control-to-output(vo(s)/d(s))


%drawing the Bode diagrams
figure(1)
bode(vo_io),grid minor,title('vo(s)/io(s)')
hold on


figure(2)
bode(vo_vg),grid minor,title('vo(s)/vg(s)')
hold on


figure(3)
bode(vo_d),grid minor,title('vo(s)/d(s)')
```

```
hold on
end
```

After running the code, the results shown in Figs. 2.21, 2.22, and 2.23 are obtained. These figures show the effect of load changes on the converter dynamics.

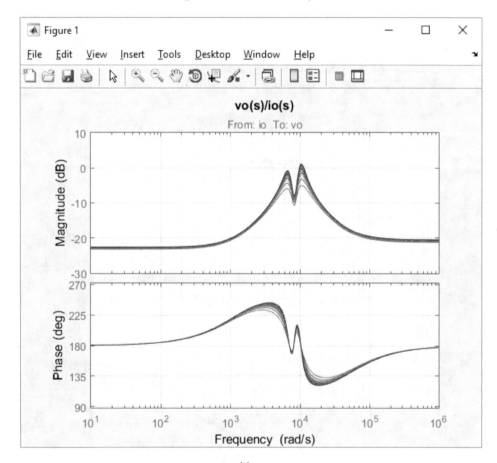

Figure 2.21: Effect of load changes on the $\frac{v_o(s)}{i_o(s)}$.

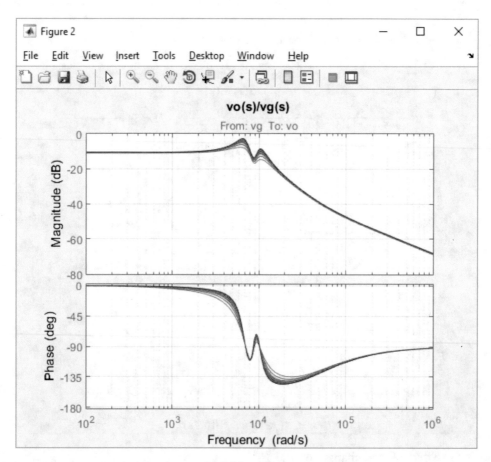

Figure 2.22: Effect of load changes on the $\frac{v_o(s)}{v_g(s)}$.

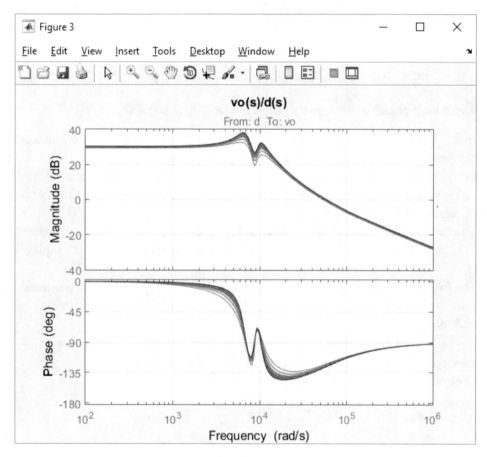

Figure 2.23: Effect of load changes on the $\frac{v_o(s)}{d(s)}$.

2.7   EXTRACTION OF ADDITIVE/MULTIPLICATIVE UNCERTAINTY MODELS

Additive and multiplicative uncertainty models are the most useful models to describe the uncertainties in DC-DC converters. This section shows how one can extract the additive/multiplicative uncertainty model of the studied Zeta converter.

If $G_p(s)$, $G_o(s)$, and I show the perturbed system dynamics, a nominal model description of the physical system and identity matrix, respectively, then:

- sdditive perturbation: $G_p(s) = G_o(s) + \Delta_{add}(s)$;

- input multiplicative perturbation: $G_p(s) = G_o(s) \times [I + \Delta_{mul}(s)]$; and

- output multiplicative perturbation: $G_p(s) = [I + \Delta_{mul}(s)] \times G_o(s)$.

In Single-Input Single-Output (SISO) systems, there is no difference between Input multiplicative perturbation and output multiplicative perturbation.

In order to extract the upper bound of additive uncertainty ($\Delta_{add}(s)$), the code must draws the $G_p(s) - G_o(s)$. In order to extract the upper bound of multiplicative uncertainty ($\Delta_{mul}(s)$), the code must draws the $\frac{G_p(s) - G_o(s)}{G_o(s)}$.

The nominal load assumed to be 6 Ω. So, the model obtained with 6 Ω load is the nominal model ($G_o(s)$). Other models (i.e., models obtained with different values of load) play the role of $G_p(s)$.

The following code draws the $G_p(s) - G_o(s)$ and $\frac{G_p(s) - G_o(s)}{G_o(s)}$. Results produced in this section are used in the next section to obtain the upper bound of the additive/multiplicative uncertainties.

```
%This program study the effect of change in load
%load changes as follows:
%
%   Rnominal=6 ohm
% 1 ohm < R < 11 ohm
%
%we want to obtain the additive and multiplicative
%uncertainty representation of the system as load
%changes
clc
clear all
close all

%Desired load range for sweep
```

```
Rnominal=6;
Rmin=1;
R_delta=.5;
Rmax=11;

VG=20;        %Value of input DC source
rg=0;         %Internal resistance of input DC source
rds=.01;      %MOSFET on resistance
C1=100e-6;    %Capacitor C1 value
C2=220e-6;    %Capacitor C2 value
rC1=.19;      %Capacitor C1 Equivalent Series Resistance(ESR)
rC2=.095;     %Capacitor C2 Equivalent Series Resistance(ESR)
L1=100e-6;    %Inductor L1 value
L2=55e-6;     %Inductor L2 value
rL1=1e-3;     %Inductor L1 Equivalent Series Resistance(ESR)
rL2=.55e-3;   %Inductor L2 Equivalent Series Resistance(ESR)
rD=.01;       %Diode series resistance
VD=.7;        %Diode voltage drop
D=.23;        %Duty cylcle
IO=0;         %Average value of output current source
fsw=100e3;    %Switching frequency

%Nominal transfer functions for R=Rnominal=6 ohm
%These results are obtained in the previous analysis.
s=tf('s');
DEN=(s^2+2239*s+4.76e7)*(s^2+2767*s+1.026e8);
vo_io_nominal=-.093519*(s+4.785e4)*(s+1163)*
   (s^2+1396*s+6.882e7)/DEN;
vo_vg_nominal=391.08*(s+4.785e4)*(s^2+1473*s+7.7e7)/DEN;
vo_d_nominal=43775*(s+4.785e4)*(s^2+1371*s+7.696e7)/DEN;
%frequency response of nominal transfer functions in the
%0.1 Rad/s - 100000 Rad/s range
%Theses frequency responses are used to calculating the
%upper bound of uncertainty weights.
omega=logspace(-1,5,200);
vo_io_nominal_frd=frd(vo_io_nominal,omega);
vo_vg_nominal_frd=frd(vo_vg_nominal,omega);
vo_d_nominal_frd=frd(vo_d_nominal,omega);
```

```matlab
n=0;
N=length([Rmin:R_delta:Rmax]);

for R=[Rmin:R_delta:Rmax]

if R==Rnominal
    continue
end

n=n+1;
disp('Percentage of work done:')
disp(n/N*100) %shows the progress of the loop

syms iL1 iL2 vC1 vC2 io vg vD d
% iL1: Inductor L1 current
% iL2: Inductor L2 current
% vC1: Capacitor C1 voltage
% vC2: Capacitor C2 voltage
% io : Output current source
% vg : Input DC source
% vD : Diode voltage drop
% d  : Duty cycle

%Closed MOSFET Equations
diL1_dt_MOSFET_close=(-(rL1+rg+rds)*iL1-(rg+rds)*iL2+vg)/L1;
diL2_dt_MOSFET_close=(-(rg+rds)*iL1-(rg+rds+rC1+rL2+R*rC2/
    (R+rC2))*iL2+vC1-R/(R+rC2)*vC2+R*rC2/(R+rC2)*io+vg)/L2;
dvC1_dt_MOSFET_close=(-iL2)/C1;
dvC2_dt_MOSFET_close=(R/(R+rC2)*iL2-1/(R+rC2)*vC2-R/
    (R+rC2)*io)/C2;
vo_MOSFET_close=R*rC2/(R+rC2)*iL2+R/(R+rC2)*vC2-R*rC2/(R+rC2)*io;

%Opened MOSFET Equations
diL1_dt_MOSFET_open=(-(rL1+rC1+rD)*iL1-rD*iL2-vC1-vD)/L1;
diL2_dt_MOSFET_open=(-rD*iL1-(rD+rL2+R*rC2/(R+rC2))*iL2-R/
    (R+rC2)*vC2+R*rC2/(R+rC2)*io-vD)/L2;
dvC1_dt_MOSFET_open=(iL1)/C1;
dvC2_dt_MOSFET_open=(R/(R+rC2)*iL2-1/(R+rC2)*vC2-R/
    (R+rC2)*io)/C2;
```

```
vo_MOSFET_open=R*rC2/(R+rC2)*iL2+R/(R+rC2)*vC2-R*rC2/(R+rC2)*io;

%Averaging
averaged_diL1_dt=simplify(d*diL1_dt_MOSFET_close+(1-d)*
    diL1_dt_MOSFET_open);
averaged_diL2_dt=simplify(d*diL2_dt_MOSFET_close+(1-d)*
    diL2_dt_MOSFET_open);
averaged_dvC1_dt=simplify(d*dvC1_dt_MOSFET_close+(1-d)*
    dvC1_dt_MOSFET_open);
averaged_dvC2_dt=simplify(d*dvC2_dt_MOSFET_close+(1-d)*
    dvC2_dt_MOSFET_open);
averaged_vo=simplify(d*vo_MOSFET_close+(1-d)*vo_MOSFET_open);

%Substituting the steady values of input DC voltage source,
%Diode voltage drop, Duty cycle and output current source
%and calculating the DC operating point
right_side_of_averaged_diL1_dt=subs(averaged_diL1_dt,
    [vg vD d io],[VG VD D IO]);
right_side_of_averaged_diL2_dt=subs(averaged_diL2_dt,
    [vg vD d io],[VG VD D IO]);
right_side_of_averaged_dvC1_dt=subs(averaged_dvC1_dt,
    [vg vD d io],[VG VD D IO]);
right_side_of_averaged_dvC2_dt=subs(averaged_dvC2_dt,
    [vg vD d io],[VG VD D IO]);

DC_OPERATING_POINT=
solve(right_side_of_averaged_diL1_dt==0,
    right_side_of_averaged_diL2_dt==0,
    right_side_of_averaged_dvC1_dt==0,
    right_side_of_averaged_dvC2_dt==0,'iL1','iL2','vC1','vC2');

IL1=eval(DC_OPERATING_POINT.iL1);
IL2=eval(DC_OPERATING_POINT.iL2);
VC1=eval(DC_OPERATING_POINT.vC1);
VC2=eval(DC_OPERATING_POINT.vC2);
VO=eval(subs(averaged_vo,[iL1 iL2 vC1 vC2 io],
    [IL1 IL2 VC1 VC2 IO]));

%Linearizing the averaged equations around the DC
```

```
%operating point. We want to obtain
%the matrix A, B, C, and D
%        .
%        x=Ax+Bu
%        y=Cx+Du
%
%where,
%        x=[iL1 iL2 vC1 vC2]'
%        u=[io vg d]'
%Since we used the variables D for steady state duty
%ratio and C to show the capacitors values we use
%AA, BB, CC, and DD instead of A, B, C, and D.

%Calculating the matrix A
A11=subs(simplify(diff(averaged_diL1_dt,iL1)),
    [iL1 iL2 vC1 vC2 d io],[IL1 IL2 VC1 VC2 D IO]);
A12=subs(simplify(diff(averaged_diL1_dt,iL2)),
    [iL1 iL2 vC1 vC2 d io],[IL1 IL2 VC1 VC2 D IO]);
A13=subs(simplify(diff(averaged_diL1_dt,vC1)),
    [iL1 iL2 vC1 vC2 d io],[IL1 IL2 VC1 VC2 D IO]);
A14=subs(simplify(diff(averaged_diL1_dt,vC2)),
    [iL1 iL2 vC1 vC2 d io],[IL1 IL2 VC1 VC2 D IO]);

A21=subs(simplify(diff(averaged_diL2_dt,iL1)),
    [iL1 iL2 vC1 vC2 d io],[IL1 IL2 VC1 VC2 D IO]);
A22=subs(simplify(diff(averaged_diL2_dt,iL2)),
    [iL1 iL2 vC1 vC2 d io],[IL1 IL2 VC1 VC2 D IO]);
A23=subs(simplify(diff(averaged_diL2_dt,vC1)),
    [iL1 iL2 vC1 vC2 d io],[IL1 IL2 VC1 VC2 D IO]);
A24=subs(simplify(diff(averaged_diL2_dt,vC2)),
    [iL1 iL2 vC1 vC2 d io],[IL1 IL2 VC1 VC2 D IO]);

A31=subs(simplify(diff(averaged_dvC1_dt,iL1)),
    [iL1 iL2 vC1 vC2 d io],[IL1 IL2 VC1 VC2 D IO]);
A32=subs(simplify(diff(averaged_dvC1_dt,iL2)),
    [iL1 iL2 vC1 vC2 d io],[IL1 IL2 VC1 VC2 D IO]);
A33=subs(simplify(diff(averaged_dvC1_dt,vC1)),
    [iL1 iL2 vC1 vC2 d io],[IL1 IL2 VC1 VC2 D IO]);
A34=subs(simplify(diff(averaged_dvC1_dt,vC2)),
```

```
  [iL1 iL2 vC1 vC2 d io],[IL1 IL2 VC1 VC2 D IO]);

A41=subs(simplify(diff(averaged_dvC2_dt,iL1)),
    [iL1 iL2 vC1 vC2 d io],[IL1 IL2 VC1 VC2 D IO]);
A42=subs(simplify(diff(averaged_dvC2_dt,iL2)),
    [iL1 iL2 vC1 vC2 d io],[IL1 IL2 VC1 VC2 D IO]);
A43=subs(simplify(diff(averaged_dvC2_dt,vC1)),
    [iL1 iL2 vC1 vC2 d io],[IL1 IL2 VC1 VC2 D IO]);
A44=subs(simplify(diff(averaged_dvC2_dt,vC2)),
    [iL1 iL2 vC1 vC2 d io],[IL1 IL2 VC1 VC2 D IO]);

AA=eval([A11 A12 A13 A14;
         A21 A22 A23 A24;
         A31 A32 A33 A34;
         A41 A42 A43 A44]);

%Calculating the matrix B
B11=subs(simplify(diff(averaged_diL1_dt,io)),
    [iL1 iL2 vC1 vC2 d vD io vg],[IL1 IL2 VC1 VC2 D VD IO VG]);
B12=subs(simplify(diff(averaged_diL1_dt,vg)),
    [iL1 iL2 vC1 vC2 d vD io vg],[IL1 IL2 VC1 VC2 D VD IO VG]);
B13=subs(simplify(diff(averaged_diL1_dt,d)),
    [iL1 iL2 vC1 vC2 d vD io vg],[IL1 IL2 VC1 VC2 D VD IO VG]);

B21=subs(simplify(diff(averaged_diL2_dt,io)),
    [iL1 iL2 vC1 vC2 d vD io vg],[IL1 IL2 VC1 VC2 D VD IO VG]);
B22=subs(simplify(diff(averaged_diL2_dt,vg)),
    [iL1 iL2 vC1 vC2 d vD io vg],[IL1 IL2 VC1 VC2 D VD IO VG]);
B23=subs(simplify(diff(averaged_diL2_dt,d)),
    [iL1 iL2 vC1 vC2 d vD io vg],[IL1 IL2 VC1 VC2 D VD IO VG]);

B31=subs(simplify(diff(averaged_dvC1_dt,io)),
    [iL1 iL2 vC1 vC2 d vD io vg],[IL1 IL2 VC1 VC2 D VD IO VG]);
B32=subs(simplify(diff(averaged_dvC1_dt,vg)),
    [iL1 iL2 vC1 vC2 d vD io vg],[IL1 IL2 VC1 VC2 D VD IO VG]);
B33=subs(simplify(diff(averaged_dvC1_dt,d)),
    [iL1 iL2 vC1 vC2 d vD io vg],[IL1 IL2 VC1 VC2 D VD IO VG]);

B41=subs(simplify(diff(averaged_dvC2_dt,io)),
```

```
    [iL1 iL2 vC1 vC2 d vD io vg],[IL1 IL2 VC1 VC2 D VD IO VG]);
B42=subs(simplify(diff(averaged_dvC2_dt,vg)),
    [iL1 iL2 vC1 vC2 d vD io vg],[IL1 IL2 VC1 VC2 D VD IO VG]);
B43=subs(simplify(diff(averaged_dvC2_dt,d)),
    [iL1 iL2 vC1 vC2 d vD io vg],[IL1 IL2 VC1 VC2 D VD IO VG]);

BB=eval([B11 B12 B13;
         B21 B22 B23;
         B31 B32 B33;
         B41 B42 B43]);

%Calculating the matrix C
C11=subs(simplify(diff(averaged_vo,iL1)),[iL1 iL2 vC1 vC2 d io],
    [IL1 IL2 VC1 VC2 D IO]);
C12=subs(simplify(diff(averaged_vo,iL2)),[iL1 iL2 vC1 vC2 d io],
    [IL1 IL2 VC1 VC2 D IO]);
C13=subs(simplify(diff(averaged_vo,vC1)),[iL1 iL2 vC1 vC2 d io],
    [IL1 IL2 VC1 VC2 D IO]);
C14=subs(simplify(diff(averaged_vo,vC2)),[iL1 iL2 vC1 vC2 d io],
    [IL1 IL2 VC1 VC2 D IO]);

CC=eval([C11 C12 C13 C14]);

D11=subs(simplify(diff(averaged_vo,io)),
    [iL1 iL2 vC1 vC2 d vD io vg],[IL1 IL2 VC1 VC2 D VD IO VG]);
D12=subs(simplify(diff(averaged_vo,vg)),
    [iL1 iL2 vC1 vC2 d vD io vg],[IL1 IL2 VC1 VC2 D VD IO VG]);
D13=subs(simplify(diff(averaged_vo,d)),
    [iL1 iL2 vC1 vC2 d vD io vg],[IL1 IL2 VC1 VC2 D VD IO VG]);

%Calculating the matrix D
DD=eval([D11 D12 D13]);

%Producing the State Space Model and obtaining the small
%signal transfer functions
sys=ss(AA,BB,CC,DD);
sys.inputname={'io';'vg';'d'};
sys.outputname={'vo'};
```

```matlab
vo_io=tf(sys(1,1)); %Output impedance transfer function
                    %vo(s)/io(s)
vo_vg=tf(sys(1,2)); %vo(s)/vg(s)
vo_d=tf(sys(1,3));  %Control-to-output(vo(s)/d(s))

%drawing the Bode diagrams

%Additive uncertainty in the vo(s)/io(s)
figure(1)
vo_io_frd=frd(vo_io,omega);
diff_vo_io=vo_io_frd-vo_io_nominal_frd;
bodemag(diff_vo_io),grid minor,title('Additive uncertainty
   in vo(s)/io(s)')
hold on

%Additive uncertainty in the vo(s)/vg(s)
figure(2)
vo_vg_frd=frd(vo_vg,omega);
diff_vo_vg=vo_vg_frd-vo_vg_nominal_frd;
bodemag(diff_vo_vg),grid minor,title('Additive uncertainty
   in vo(s)/vg(s)')
hold on

%Additive uncertainty in the vo(s)/d(s)
figure(3)
vo_d_frd=frd(vo_d,omega);
diff_vo_d=vo_d_frd-vo_d_nominal_frd;
bodemag(diff_vo_d),grid minor,title('Additive uncertainty
   in vo(s)/d(s)')
hold on

%Multiplicative uncertainty in the vo(s)/io(s)
figure(4)
vo_io_frd=frd(vo_io,omega);
diff_vo_io=vo_io_frd-vo_io_nominal_frd;
bodemag(diff_vo_io/vo_io_nominal_frd),grid minor,
   title('Multipicative uncertainty in vo(s)/io(s)')
hold on
```

```
%Multiplicative uncertainty in the vo(s)/vg(s)
figure(5)
vo_vg_frd=frd(vo_vg,omega);
diff_vo_vg=vo_vg_frd-vo_vg_nominal_frd;
bodemag(diff_vo_vg/vo_vg_nominal_frd),grid minor,
    title('Multipicative uncertainty in vo(s)/vg(s)')
hold on

%Multiplicative uncertainty in the vo(s)/d(s)
figure(6)
vo_d_frd=frd(vo_d,omega);
diff_vo_d=vo_d_frd-vo_d_nominal_frd;
bodemag(diff_vo_d/vo_d_nominal_frd),grid minor,
    title('Multipicative uncertainty vo(s)/d(s)')
hold on
end
```

After running the code, the results shown in Figs. 2.24, 2.25, and 2.26 are obtained.

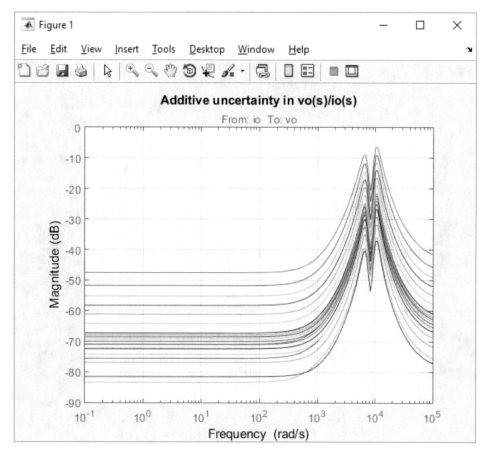

Figure 2.24: Additive uncertainty in the $\frac{v_o(s)}{i_o(s)}$.

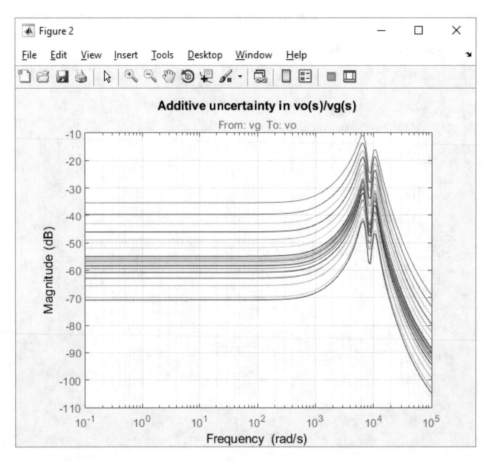

Figure 2.25: Additive uncertainty in the $\frac{v_o(s)}{v_g(s)}$.

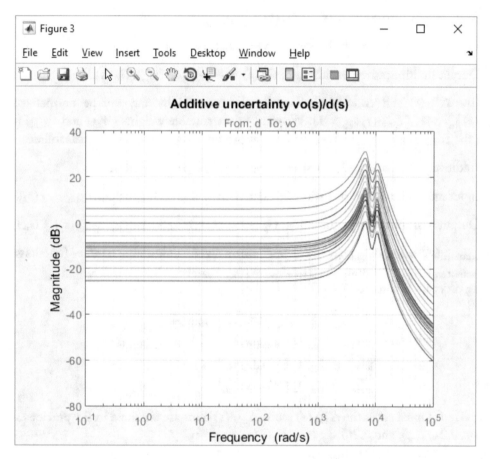

Figure 2.26: Additive uncertainty in the $\frac{v_o(s)}{d(s)}$.

2.8   UPPER BOUND OF ADDITIVE/MULTIPLICATIVE UNCERTAINTY MODELS

Although results of previous analysis shows the effect of load variations on the converters dynamics graphically, it gives no information about the suitable weights to model the uncertainty. In the previous section, we used the following notations.

- Additive perturbation: $G_p(s) = G_o(s) + \Delta_{\text{add}}(s)$.

- Input multiplicative perturbation: $G_p(s) = G_o(s) \times [I + \Delta_{\text{mul}}(s)]$.

- Output multiplicative perturbation: $G_p(s) = [I + \Delta_{\text{mul}}(s)] \times G_o(s)$.

The $\Delta_{\text{add}}(s)$ and $\Delta_{\text{mul}}(s)$ used in the above notations may not be normalized (i.e., $\|\Delta_{\text{add}}(s)\|_\infty > 1$, $\|\Delta_{\text{mul}}(s)\|_\infty > 1$). We want to use suitable weights (W_{add} and W_{mul}) to normalize the $\Delta_{\text{add}}(s)$ and $\Delta_{\text{mul}}(s)$. So, we want to modify the previous notation as follows.

- Additive perturbation: $G_p(s) = G_o(s) + W_{\text{add}}(s)\Delta_{\text{add,normalized}}(s)$.

- Input multiplicative perturbation: $G_p(s) = G_o(s) \times [I + W_{\text{mul}}(s)\Delta_{\text{mul,normalized}}(s)]$.

- Output multiplicative perturbation: $G_p(s) = [I + W_{\text{mul}}(s)\Delta_{\text{mul,normalized}}(s)] \times G_o(s)$.

$\Delta_{\text{add,normalized}}(s)$ and $\Delta_{\text{mul,normalized}}(s)$ are normalized unknown tranfer functions, i.e., $\|\Delta_{\text{add,normalized}}(s)\|_\infty < 1$ and $\|\Delta_{\text{mul,normalized}}(s)\|_\infty < 1$.

In order to obtain the $W_{\text{add}}(s)$,

$$\|\Delta_{\text{add}}(s)\|_\infty \leq \|W_{\text{add}}(s)\Delta_{\text{add,normalized}}(s)\|_\infty$$
$$\implies \|\Delta_{\text{add}}(s)\|_\infty \leq \|W_{\text{add}}(s)\|_\infty \times \|\Delta_{\text{add,normalized}}(s)\|_\infty$$
$$\implies \|\Delta_{\text{add}}(s)\|_\infty \leq \|W_{\text{add}}(s)\|_\infty \times 1$$
$$\implies \|\Delta_{\text{add}}(s)\|_\infty \leq \|W_{\text{add}}(s)\|_\infty.$$

So, the $W_{\text{add}}(s)$ must cover the $\Delta_{\text{add}}(s)$ plots. $\Delta_{\text{add}}(s)$ plots are obtained in the previous section (see Figs. 2.24, 2.25, and 2.26).

In order to obtain the $W_{\text{mul}}(s)$,

$$\|\Delta_{\text{mul}}(s)\|_\infty \leq \|W_{\text{mul}}(s)\Delta_{\text{mul,normalized}}(s)\|_\infty$$
$$\implies \|\Delta_{\text{mul}}(s)\|_\infty \leq \|W_{\text{mul}}(s)\|_\infty \times \|\Delta_{\text{mul,normalized}}(s)\|_\infty$$
$$\implies \|\Delta_{\text{mul}}(s)\|_\infty \leq \|W_{\text{mul}}(s)\|_\infty \times 1$$
$$\implies \|\Delta_{\text{mul}}(s)\|_\infty \leq \|W_{\text{mul}}(s)\|_\infty.$$

So, the $W_{\text{mul}}(s)$ must cover the $\Delta_{\text{mul}}(s)$ plots. $\Delta_{\text{mul}}(s)$ plots are obtained in the previous section (see Figs. 2.27, 2.28, and 2.29).

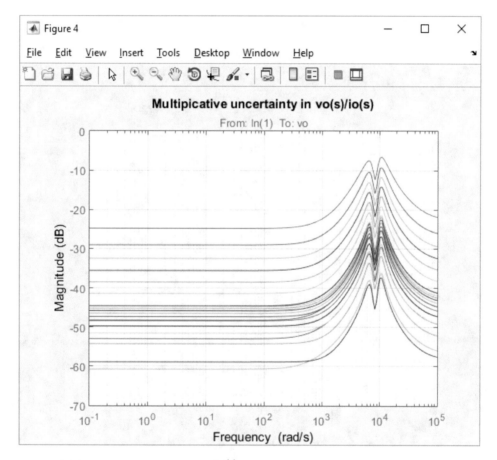

Figure 2.27: Additive uncertainty in the $\frac{v_o(s)}{i_o(s)}$.

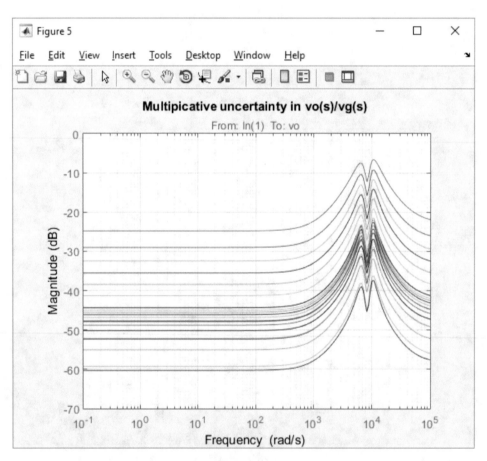

Figure 2.28: Additive uncertainty in the $\frac{v_o(s)}{v_g(s)}$.

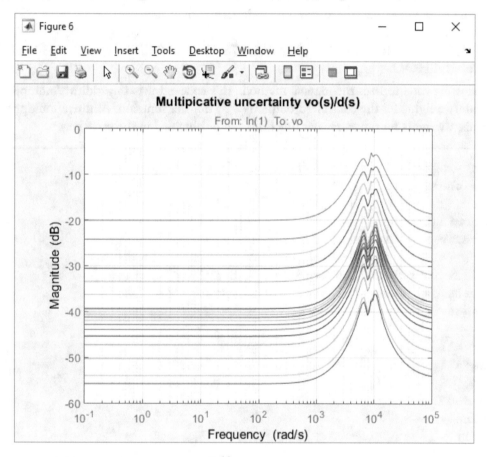

Figure 2.29: Additive uncertainty in the $\frac{v_o(s)}{d(s)}$.

One can obtain the weights ($W_{add}(s)$ and $W_{mul}(s)$) in two ways.

- **Manual method:** In this method, the user selects some of the points lie above the obtained transfer functions ($\Delta_{add}(s)$ and $\Delta_{mul}(s)$ plots). MATLAB® fits suitable weights to the selected points using fitmagfrd command.

- **Using the "ucover" command:** MATLAB® has a built in function named ucover which can be used to extract the uncertainty weights automatically.

2.8.1 EXTRACTION OF UNCERTAINTY WEIGHTS USING THE MANUAL METHOD

The following code realizes the manual method. This code extracts the additive/multiplicative uncertainty weights for the control-to-output ($\frac{v_o(s)}{d(s)}$) transfer function. Additive/multiplicative uncertainty weights for other transfer functions can be obtained in the same way.

```
%This program study the effect of change in load
%load changes as follows:
%
%   Rnominal=6 ohm
% 1 ohm < R < 11 ohm
%
%we want to obtain the additive and multiplicative
%uncertainty representation of the system as load
%changes
clc
clear all
close all

%Desired load range for sweep
Rnominal=6;
Rmin=1;
R_delta=.5;
Rmax=11;

VG=20;       %Value of input DC source
rg=0;        %Internal resistance of input DC source
rds=.01;     %MOSFET on resistance
C1=100e-6;   %Capacitor C1 value
C2=220e-6;   %Capacitor C2 value
rC1=.19;     %Capacitor C1 Equivalent Series Resistance(ESR)
```

```
rC2=.095;    %Capacitor C2 Equivalent Series Resistance(ESR)
L1=100e-6;   %Inductor L1 value
L2=55e-6;    %Inductor L2 value
rL1=1e-3;    %Inductor L1 Equivalent Series Resistance(ESR)
rL2=.55e-3;  %Inductor L2 Equivalent Series Resistance(ESR)
rD=.01;      %Diode series resistance
VD=.7;       %Diode voltage drop
D=.23;       %Duty cylcle
IO=0;        %Average value of output current source
fsw=100e3;   %Switching frequency

%Nominal transfer functions for R=Rnominal=6 ohm
%These results are obtained in the previous analysis.
s=tf('s');
DEN=(s^2+2239*s+4.76e7)*(s^2+2767*s+1.026e8);
vo_io_nominal=-.093519*(s+4.785e4)*(s+1163)*
   (s^2+1396*s+6.882e7)/DEN;
vo_vg_nominal=391.08*(s+4.785e4)*(s^2+1473*s+7.7e7)/DEN;
vo_d_nominal=43775*(s+4.785e4)*(s^2+1371*s+7.696e7)/DEN;
%frequency response of nominal transfer functions in the
%0.1 Rad/s - 100000 Rad/s range
%Theses frequency responses are used to calculating
%the upper bound of uncertainty weights.
omega=logspace(-1,5,200);
vo_io_nominal_frd=frd(vo_io_nominal,omega);
vo_vg_nominal_frd=frd(vo_vg_nominal,omega);
vo_d_nominal_frd=frd(vo_d_nominal,omega);

n=0;
N=length([Rmin:R_delta:Rmax]);

for R=[Rmin:R_delta:Rmax]

if R==Rnominal
    continue
end

n=n+1;
disp('Percentage of work done:')
```

```
disp(n/N*100) %shows the progress of the loop

syms iL1 iL2 vC1 vC2 io vg vD d
% iL1: Inductor L1 current
% iL2: Inductor L2 current
% vC1: Capacitor C1 voltage
% vC2: Capacitor C2 voltage
% io : Output current source
% vg : Input DC source
% vD : Diode voltage drop
% d  : Duty cycle

%Closed MOSFET Equations
diL1_dt_MOSFET_close=(-(rL1+rg+rds)*iL1-(rg+rds)*iL2+vg)/L1;
diL2_dt_MOSFET_close=(-(rg+rds)*iL1-(rg+rds+rC1+rL2+R*rC2/
    (R+rC2))*iL2+vC1-R/(R+rC2)*vC2+R*rC2/(R+rC2)*io+vg)/L2;
dvC1_dt_MOSFET_close=(-iL2)/C1;
dvC2_dt_MOSFET_close=(R/(R+rC2)*iL2-1/(R+rC2)*vC2-R/
    (R+rC2)*io)/C2;
vo_MOSFET_close=R*rC2/(R+rC2)*iL2+R/(R+rC2)*vC2-R*rC2/(R+rC2)*io;

%Opened MOSFET Equations
diL1_dt_MOSFET_open=(-(rL1+rC1+rD)*iL1-rD*iL2-vC1-vD)/L1;
diL2_dt_MOSFET_open=(-rD*iL1-(rD+rL2+R*rC2/(R+rC2))*iL2-R/
    (R+rC2)*vC2+R*rC2/(R+rC2)*io-vD)/L2;
dvC1_dt_MOSFET_open=(iL1)/C1;
dvC2_dt_MOSFET_open=(R/(R+rC2)*iL2-1/(R+rC2)*vC2-R/
    (R+rC2)*io)/C2;
vo_MOSFET_open=R*rC2/(R+rC2)*iL2+R/(R+rC2)*vC2-R*rC2/(R+rC2)*io;

%Averaging
averaged_diL1_dt=simplify(d*diL1_dt_MOSFET_close+(1-d)*
    diL1_dt_MOSFET_open);
averaged_diL2_dt=simplify(d*diL2_dt_MOSFET_close+(1-d)*
    diL2_dt_MOSFET_open);
averaged_dvC1_dt=simplify(d*dvC1_dt_MOSFET_close+(1-d)*
    dvC1_dt_MOSFET_open);
averaged_dvC2_dt=simplify(d*dvC2_dt_MOSFET_close+(1-d)*
    dvC2_dt_MOSFET_open);
```

```
averaged_vo=simplify(d*vo_MOSFET_close+(1-d)*vo_MOSFET_open);

%Substituting the steady values of input DC voltage source,
%Diode voltage drop, Duty cycle and output current source
%and calculating the DC operating point
right_side_of_averaged_diL1_dt=subs(averaged_diL1_dt,
    [vg vD d io],[VG VD D IO]);
right_side_of_averaged_diL2_dt=subs(averaged_diL2_dt,
    [vg vD d io],[VG VD D IO]);
right_side_of_averaged_dvC1_dt=subs(averaged_dvC1_dt,
    [vg vD d io],[VG VD D IO]);
right_side_of_averaged_dvC2_dt=subs(averaged_dvC2_dt,
    [vg vD d io],[VG VD D IO]);

DC_OPERATING_POINT=
solve(right_side_of_averaged_diL1_dt==0,
    right_side_of_averaged_diL2_dt==0,
    right_side_of_averaged_dvC1_dt==0,
    right_side_of_averaged_dvC2_dt==0,'iL1','iL2','vC1','vC2');

IL1=eval(DC_OPERATING_POINT.iL1);
IL2=eval(DC_OPERATING_POINT.iL2);
VC1=eval(DC_OPERATING_POINT.vC1);
VC2=eval(DC_OPERATING_POINT.vC2);
VO=eval(subs(averaged_vo,[iL1 iL2 vC1 vC2 io],
    [IL1 IL2 VC1 VC2 IO]));

%Linearizing the averaged equations around the DC
%operating point. We want to obtain the matrix A, B, C, and D
%         .
%       x=Ax+Bu
%       y=Cx+Du
%
%where,
%       x=[iL1 iL2 vC1 vC2]'
%       u=[io vg d]'
%Since we used the variables D for steady state duty
%ratio and C to show the capacitors values we use AA,
%BB, CC, and DD instead of A, B, C, and D.
```

```
%Calculating the matrix A
A11=subs(simplify(diff(averaged_diL1_dt,iL1)),
    [iL1 iL2 vC1 vC2 d io],[IL1 IL2 VC1 VC2 D IO]);
A12=subs(simplify(diff(averaged_diL1_dt,iL2)),
    [iL1 iL2 vC1 vC2 d io],[IL1 IL2 VC1 VC2 D IO]);
A13=subs(simplify(diff(averaged_diL1_dt,vC1)),
    [iL1 iL2 vC1 vC2 d io],[IL1 IL2 VC1 VC2 D IO]);
A14=subs(simplify(diff(averaged_diL1_dt,vC2)),
    [iL1 iL2 vC1 vC2 d io],[IL1 IL2 VC1 VC2 D IO]);

A21=subs(simplify(diff(averaged_diL2_dt,iL1)),
    [iL1 iL2 vC1 vC2 d io],[IL1 IL2 VC1 VC2 D IO]);
A22=subs(simplify(diff(averaged_diL2_dt,iL2)),
    [iL1 iL2 vC1 vC2 d io],[IL1 IL2 VC1 VC2 D IO]);
A23=subs(simplify(diff(averaged_diL2_dt,vC1)),
    [iL1 iL2 vC1 vC2 d io],[IL1 IL2 VC1 VC2 D IO]);
A24=subs(simplify(diff(averaged_diL2_dt,vC2)),
    [iL1 iL2 vC1 vC2 d io],[IL1 IL2 VC1 VC2 D IO]);

A31=subs(simplify(diff(averaged_dvC1_dt,iL1)),
    [iL1 iL2 vC1 vC2 d io],[IL1 IL2 VC1 VC2 D IO]);
A32=subs(simplify(diff(averaged_dvC1_dt,iL2)),
    [iL1 iL2 vC1 vC2 d io],[IL1 IL2 VC1 VC2 D IO]);
A33=subs(simplify(diff(averaged_dvC1_dt,vC1)),
    [iL1 iL2 vC1 vC2 d io],[IL1 IL2 VC1 VC2 D IO]);
A34=subs(simplify(diff(averaged_dvC1_dt,vC2)),
    [iL1 iL2 vC1 vC2 d io],[IL1 IL2 VC1 VC2 D IO]);

A41=subs(simplify(diff(averaged_dvC2_dt,iL1)),
    [iL1 iL2 vC1 vC2 d io],[IL1 IL2 VC1 VC2 D IO]);
A42=subs(simplify(diff(averaged_dvC2_dt,iL2)),
    [iL1 iL2 vC1 vC2 d io],[IL1 IL2 VC1 VC2 D IO]);
A43=subs(simplify(diff(averaged_dvC2_dt,vC1)),
    [iL1 iL2 vC1 vC2 d io],[IL1 IL2 VC1 VC2 D IO]);
A44=subs(simplify(diff(averaged_dvC2_dt,vC2)),
    [iL1 iL2 vC1 vC2 d io],[IL1 IL2 VC1 VC2 D IO]);

AA=eval([A11 A12 A13 A14;
```

```
            A21 A22 A23 A24;
            A31 A32 A33 A34;
            A41 A42 A43 A44]);

%Calculating the matrix B
B11=subs(simplify(diff(averaged_diL1_dt,io)),
    [iL1 iL2 vC1 vC2 d vD io vg],[IL1 IL2 VC1 VC2 D VD IO VG]);
B12=subs(simplify(diff(averaged_diL1_dt,vg)),
    [iL1 iL2 vC1 vC2 d vD io vg],[IL1 IL2 VC1 VC2 D VD IO VG]);
B13=subs(simplify(diff(averaged_diL1_dt,d)),
    [iL1 iL2 vC1 vC2 d vD io vg],[IL1 IL2 VC1 VC2 D VD IO VG]);

B21=subs(simplify(diff(averaged_diL2_dt,io)),
    [iL1 iL2 vC1 vC2 d vD io vg],[IL1 IL2 VC1 VC2 D VD IO VG]);
B22=subs(simplify(diff(averaged_diL2_dt,vg)),
    [iL1 iL2 vC1 vC2 d vD io vg],[IL1 IL2 VC1 VC2 D VD IO VG]);
B23=subs(simplify(diff(averaged_diL2_dt,d)),
    [iL1 iL2 vC1 vC2 d vD io vg],[IL1 IL2 VC1 VC2 D VD IO VG]);

B31=subs(simplify(diff(averaged_dvC1_dt,io)),
    [iL1 iL2 vC1 vC2 d vD io vg],[IL1 IL2 VC1 VC2 D VD IO VG]);
B32=subs(simplify(diff(averaged_dvC1_dt,vg)),
    [iL1 iL2 vC1 vC2 d vD io vg],[IL1 IL2 VC1 VC2 D VD IO VG]);
B33=subs(simplify(diff(averaged_dvC1_dt,d)),
    [iL1 iL2 vC1 vC2 d vD io vg],[IL1 IL2 VC1 VC2 D VD IO VG]);

B41=subs(simplify(diff(averaged_dvC2_dt,io)),
    [iL1 iL2 vC1 vC2 d vD io vg],[IL1 IL2 VC1 VC2 D VD IO VG]);
B42=subs(simplify(diff(averaged_dvC2_dt,vg)),
    [iL1 iL2 vC1 vC2 d vD io vg],[IL1 IL2 VC1 VC2 D VD IO VG]);
B43=subs(simplify(diff(averaged_dvC2_dt,d)),
    [iL1 iL2 vC1 vC2 d vD io vg],[IL1 IL2 VC1 VC2 D VD IO VG]);

BB=eval([B11 B12 B13;
         B21 B22 B23;
         B31 B32 B33;
         B41 B42 B43]);

%Calculating the matrix C
```

```
C11=subs(simplify(diff(averaged_vo,iL1)),[iL1 iL2 vC1 vC2 d io],
    [IL1 IL2 VC1 VC2 D IO]);
C12=subs(simplify(diff(averaged_vo,iL2)),[iL1 iL2 vC1 vC2 d io],
    [IL1 IL2 VC1 VC2 D IO]);
C13=subs(simplify(diff(averaged_vo,vC1)),[iL1 iL2 vC1 vC2 d io],
    [IL1 IL2 VC1 VC2 D IO]);
C14=subs(simplify(diff(averaged_vo,vC2)),[iL1 iL2 vC1 vC2 d io],
    [IL1 IL2 VC1 VC2 D IO]);

CC=eval([C11 C12 C13 C14]);

D11=subs(simplify(diff(averaged_vo,io)),
    [iL1 iL2 vC1 vC2 d vD io vg],[IL1 IL2 VC1 VC2 D VD IO VG]);
D12=subs(simplify(diff(averaged_vo,vg)),
    [iL1 iL2 vC1 vC2 d vD io vg],[IL1 IL2 VC1 VC2 D VD IO VG]);
D13=subs(simplify(diff(averaged_vo,d)),
    [iL1 iL2 vC1 vC2 d vD io vg],[IL1 IL2 VC1 VC2 D VD IO VG]);

%Calculating the matrix D
DD=eval([D11 D12 D13]);

%Producing the State Space Model and obtaining the small
%signal transfer functions
sys=ss(AA,BB,CC,DD);
sys.inputname={'io';'vg';'d'};
sys.outputname={'vo'};

vo_io=tf(sys(1,1)); %Output impedance transfer function
                    %vo(s)/io(s)
vo_vg=tf(sys(1,2)); %vo(s)/vg(s)
vo_d=tf(sys(1,3));  %Control-to-output(vo(s)/d(s))

%drawing the Bode diagrams

%Additive uncertainty in the vo(s)/io(s)
figure(1)
vo_io_frd=frd(vo_io,omega);
diff_vo_io=vo_io_frd-vo_io_nominal_frd;
bodemag(diff_vo_io),grid minor,title('Additive uncertainty
```

```
   in vo(s)/io(s)')
hold on

%Additive uncertainty in the vo(s)/vg(s)
figure(2)
vo_vg_frd=frd(vo_vg,omega);
diff_vo_vg=vo_vg_frd-vo_vg_nominal_frd;
bodemag(diff_vo_vg),grid minor,title('Additive uncertainty
   in vo(s)/vg(s)')
hold on

%Additive uncertainty in the vo(s)/d(s)
figure(3)
vo_d_frd=frd(vo_d,omega);
diff_vo_d=vo_d_frd-vo_d_nominal_frd;
bodemag(diff_vo_d),grid minor,title('Additive uncertainty
   in vo(s)/d(s)')
hold on

%Multiplicative uncertainty in the vo(s)/io(s)
figure(4)
vo_io_frd=frd(vo_io,omega);
diff_vo_io=vo_io_frd-vo_io_nominal_frd;
bodemag(diff_vo_io/vo_io_nominal_frd),grid minor,
   title('Multipicative uncertainty in vo(s)/io(s)')
hold on

%Multiplicative uncertainty in the vo(s)/vg(s)
figure(5)
vo_vg_frd=frd(vo_vg,omega);
diff_vo_vg=vo_vg_frd-vo_vg_nominal_frd;
bodemag(diff_vo_vg/vo_vg_nominal_frd),grid minor,
   title('Multipicative uncertainty in vo(s)/vg(s)')
hold on

%Multiplicative uncertainty in the vo(s)/d(s)
figure(6)
vo_d_frd=frd(vo_d,omega);
diff_vo_d=vo_d_frd-vo_d_nominal_frd;
```

```matlab
bodemag(diff_vo_d/vo_d_nominal_frd),grid minor,
    title('Multipicative uncertainty vo(s)/d(s)')
hold on
end

%This code extract the upper bound for multiplicative
%uncertainty.
figure(6) % Note that Figure 6 contains the the
%multiplicative uncertainity.
%Following 2 lines of code permits the user to selects
%20 points from the Figure 6
Number_of_points=20;
[freq,resp_dB]=ginput(Number_of_points);
%Reading are in desibel(dB). The following loop find
%the magnitudes.
for i=1:Number_of_points
    resp(i)=10^(resp_dB(i)/20);
end
selected_points_frd=frd(resp,freq); %Making the frd object.
ord=2; %Order of produced weight
W=fitmagfrd(selected_points_frd,ord); %Fitting a transfer
        %function to the selected data points.
WtfMultiplicative=tf(W);
bode(WtfMultiplicative,'r--')

%This code extract the upper bound for additive uncertainty.
figure(3) %Note that Figure 3 contains the the additive
%uncertainity.
%Following 2 lines of code permits the user to selects
%20 points from the Figure 3
Number_of_points=20;
[freq,resp_dB]=ginput(Number_of_points);
%Reading are in desibel(dB). The following loop find
%the magnitudes.
for i=1:Number_of_points
    resp(i)=10^(resp_dB(i)/20);
end
selected_points_frd=frd(resp,freq);    %Making the frd object.
ord=3; %we used a 3rd order transfer function to model the
```

```
        %upper bound of uncertainty.
W=fitmagfrd(selected_points_frd,ord); %Fitting a transfer
        %function to the selected data points.
WtfAdditive=tf(W);
bode(WtfAdditive,'g--')
```

After running the code, the user must selects the 20 points which lies above all the drawn transfer functions. Analysis results are shown in Figs. 2.30–2.33. Obtained fitted weights are shown with a dashed line.

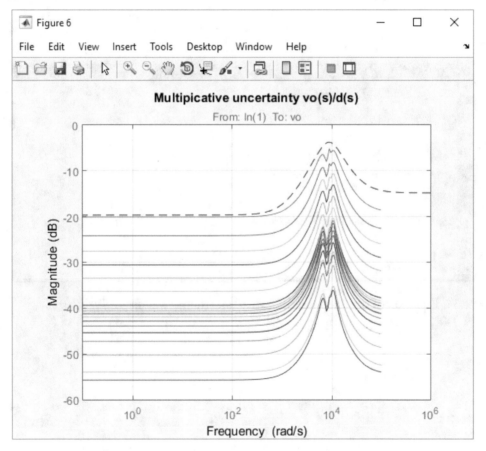

Figure 2.30: Obtaining the multiplicative uncertainty weight for $\frac{v_o(s)}{d(s)}$. The obtained weight is shown with a dashed line.

```
Command Window                                              ⊙

  >> zpk(WtfMultiplicative)

  ans =

    0.18113 (s+3.556e04) (s+1194)
    -----------------------------
      (s^2 + 1.026e04s + 7.38e07)

  Continuous-time zero/pole/gain model.

fx >> |
```

Figure 2.31: Equation of obtained multiplicative uncertainty weight for $\frac{v_o(s)}{d(s)}$.

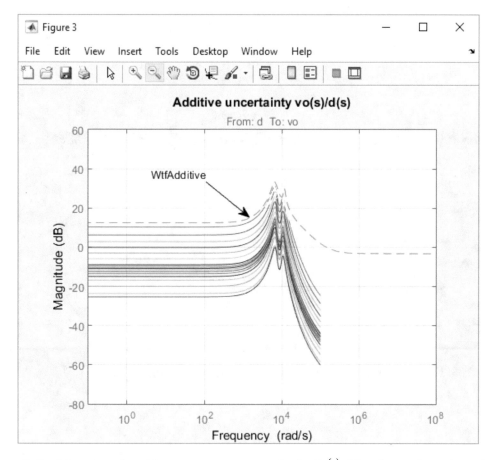

Figure 2.32: Obtaining the additive uncertainty weight for $\frac{v_o(s)}{d(s)}$. The obtained weight is shown with a dashed line.

```
Command Window                                                    ⊙

  >> zpk(WtfAdditive)

  ans =

    0.68713 (s+1.473e05) (s+1851) (s^2 + 3640s + 1.211e08)
    ------------------------------------------------------
       (s^2 + 2031s + 4.056e07) (s^2 + 1480s + 1.324e08)

  Continuous-time zero/pole/gain model.

fx >> |
```

Figure 2.33: Equation of obtained additive uncertainty weight for $\frac{v_o(s)}{d(s)}$.

2.8.2   EXTRACTION OF UNCERTAINTY WEIGHTS USING THE MATLAB®UCOVER COMMAND

The following code uses the ucover command to extract the uncertainty weights. This code extracts the multiplicative uncertainty weights for the control-to-output $\left(\frac{v_o(s)}{d(s)}\right)$ transfer function. Multiplicative uncertainty weights for other transfer functions can be obtained in the same way.

```
%This program study the effect of change in load
%load changes as follows:
%
%  Rnominal=6 ohm
% 1 ohm < R < 11 ohm
%
%we want to obtain the multiplicative
%uncertainty representation of the vo(s)/d(s)
%We use the "ucover" command here.
clc
clear all
close all

%Desired load range for sweep
Rnominal=6;
Rmin=1;
R_delta=.5;
Rmax=11;

VG=20;          %Value of input DC source
rg=0;           %Internal resistance of input DC source
rds=.01;        %MOSFET on resistance
C1=100e-6;      %Capacitor C1 value
C2=220e-6;      %Capacitor C2 value
rC1=.19;        %Capacitor C1 Equivalent Series Resistance(ESR)
rC2=.095;       %Capacitor C2 Equivalent Series Resistance(ESR)
L1=100e-6;      %Inductor L1 value
L2=55e-6;       %Inductor L2 value
rL1=1e-3;       %Inductor L1 Equivalent Series Resistance(ESR)
rL2=.55e-3;     %Inductor L2 Equivalent Series Resistance(ESR)
rD=.01;         %Diode series resistance
VD=.7;          %Diode voltage drop
D=.23;          %Duty cylcle
```

```
I0=0;         %Average value of output current source
fsw=100e3;    %Switching frequency

%Nominal transfer functions for R=Rnominal=6 ohm
%These results are obtained in the previous analysis.
s=tf('s');
DEN=(s^2+2239*s+4.76e7)*(s^2+2767*s+1.026e8);
vo_io_nominal=-.093519*(s+4.785e4)*(s+1163)*
    (s^2+1396*s+6.882e7)/DEN;
vo_vg_nominal=391.08*(s+4.785e4)*(s^2+1473*s+7.7e7)/DEN;
vo_d_nominal=43775*(s+4.785e4)*(s^2+1371*s+7.696e7)/DEN;

n=0;
N=length([Rmin:R_delta:Rmax]);

for R=[Rmin:R_delta:Rmax]

if R==Rnominal
    continue
end

n=n+1;
disp('Percentage of work done:')
disp(n/N*100) %shows the progress of the loop

syms iL1 iL2 vC1 vC2 io vg vD d
% iL1: Inductor L1 current
% iL2: Inductor L2 current
% vC1: Capacitor C1 voltage
% vC2: Capacitor C2 voltage
% io : Output current source
% vg : Input DC source
% vD : Diode voltage drop
% d  : Duty cycle

%Closed MOSFET Equations
diL1_dt_MOSFET_close=(-(rL1+rg+rds)*iL1-(rg+rds)*iL2+vg)/L1;
diL2_dt_MOSFET_close=(-(rg+rds)*iL1-(rg+rds+rC1+rL2+R*rC2/
    (R+rC2))*iL2+vC1-R/(R+rC2)*vC2+R*rC2/(R+rC2)*io+vg)/L2;
```

```
dvC1_dt_MOSFET_close=(-iL2)/C1;
dvC2_dt_MOSFET_close=(R/(R+rC2)*iL2-1/(R+rC2)*vC2-R/
    (R+rC2)*io)/C2;
vo_MOSFET_close=R*rC2/(R+rC2)*iL2+R/(R+rC2)*vC2-R*rC2/(R+rC2)*io;

%Opened MOSFET Equations
diL1_dt_MOSFET_open=(-(rL1+rC1+rD)*iL1-rD*iL2-vC1-vD)/L1;
diL2_dt_MOSFET_open=(-rD*iL1-(rD+rL2+R*rC2/(R+rC2))*iL2-R/
    (R+rC2)*vC2+R*rC2/(R+rC2)*io-vD)/L2;
dvC1_dt_MOSFET_open=(iL1)/C1;
dvC2_dt_MOSFET_open=(R/(R+rC2)*iL2-1/(R+rC2)*vC2-R/
    (R+rC2)*io)/C2;
vo_MOSFET_open=R*rC2/(R+rC2)*iL2+R/(R+rC2)*vC2-R*rC2/(R+rC2)*io;

%Averaging
averaged_diL1_dt=simplify(d*diL1_dt_MOSFET_close+(1-d)*
    diL1_dt_MOSFET_open);
averaged_diL2_dt=simplify(d*diL2_dt_MOSFET_close+(1-d)*
    diL2_dt_MOSFET_open);
averaged_dvC1_dt=simplify(d*dvC1_dt_MOSFET_close+(1-d)*
    dvC1_dt_MOSFET_open);
averaged_dvC2_dt=simplify(d*dvC2_dt_MOSFET_close+(1-d)*
    dvC2_dt_MOSFET_open);
averaged_vo=simplify(d*vo_MOSFET_close+(1-d)*vo_MOSFET_open);

%Substituting the steady values of input DC voltage source,
%Diode voltage drop, Duty cycle and output current source
%and calculating the DC operating point
right_side_of_averaged_diL1_dt=subs(averaged_diL1_dt,
    [vg vD d io],[VG VD D IO]);
right_side_of_averaged_diL2_dt=subs(averaged_diL2_dt,
    [vg vD d io],[VG VD D IO]);
right_side_of_averaged_dvC1_dt=subs(averaged_dvC1_dt,
    [vg vD d io],[VG VD D IO]);
right_side_of_averaged_dvC2_dt=subs(averaged_dvC2_dt,
    [vg vD d io],[VG VD D IO]);

DC_OPERATING_POINT=
solve(right_side_of_averaged_diL1_dt==0,
```

```
    right_side_of_averaged_diL2_dt==0,
    right_side_of_averaged_dvC1_dt==0,
    right_side_of_averaged_dvC2_dt==0,'iL1','iL2','vC1','vC2');

IL1=eval(DC_OPERATING_POINT.iL1);
IL2=eval(DC_OPERATING_POINT.iL2);
VC1=eval(DC_OPERATING_POINT.vC1);
VC2=eval(DC_OPERATING_POINT.vC2);
VO=eval(subs(averaged_vo,[iL1 iL2 vC1 vC2 io],
    [IL1 IL2 VC1 VC2 IO]));

%Linearizing the averaged equations around the DC operating
%point. We want to obtain the matrix A, B, C, and D
%         .
%        x=Ax+Bu
%        y=Cx+Du
%
%where,
%        x=[iL1 iL2 vC1 vC2]'
%        u=[io vg d]'
%Since we used the variables D for steady state duty
%ratio and C to show the capacitors values we use AA,
%BB, CC and DD, instead of A, B, C, and D.

%Calculating the matrix A
A11=subs(simplify(diff(averaged_diL1_dt,iL1)),
    [iL1 iL2 vC1 vC2 d io],[IL1 IL2 VC1 VC2 D IO]);
A12=subs(simplify(diff(averaged_diL1_dt,iL2)),
    [iL1 iL2 vC1 vC2 d io],[IL1 IL2 VC1 VC2 D IO]);
A13=subs(simplify(diff(averaged_diL1_dt,vC1)),
    [iL1 iL2 vC1 vC2 d io],[IL1 IL2 VC1 VC2 D IO]);
A14=subs(simplify(diff(averaged_diL1_dt,vC2)),
    [iL1 iL2 vC1 vC2 d io],[IL1 IL2 VC1 VC2 D IO]);

A21=subs(simplify(diff(averaged_diL2_dt,iL1)),
    [iL1 iL2 vC1 vC2 d io],[IL1 IL2 VC1 VC2 D IO]);
A22=subs(simplify(diff(averaged_diL2_dt,iL2)),
    [iL1 iL2 vC1 vC2 d io],[IL1 IL2 VC1 VC2 D IO]);
A23=subs(simplify(diff(averaged_diL2_dt,vC1)),
```

```
        [iL1 iL2 vC1 vC2 d io],[IL1 IL2 VC1 VC2 D IO]);
A24=subs(simplify(diff(averaged_diL2_dt,vC2)),
    [iL1 iL2 vC1 vC2 d io],[IL1 IL2 VC1 VC2 D IO]);

A31=subs(simplify(diff(averaged_dvC1_dt,iL1)),
    [iL1 iL2 vC1 vC2 d io],[IL1 IL2 VC1 VC2 D IO]);
A32=subs(simplify(diff(averaged_dvC1_dt,iL2)),
    [iL1 iL2 vC1 vC2 d io],[IL1 IL2 VC1 VC2 D IO]);
A33=subs(simplify(diff(averaged_dvC1_dt,vC1)),
    [iL1 iL2 vC1 vC2 d io],[IL1 IL2 VC1 VC2 D IO]);
A34=subs(simplify(diff(averaged_dvC1_dt,vC2)),
    [iL1 iL2 vC1 vC2 d io],[IL1 IL2 VC1 VC2 D IO]);

A41=subs(simplify(diff(averaged_dvC2_dt,iL1)),
    [iL1 iL2 vC1 vC2 d io],[IL1 IL2 VC1 VC2 D IO]);
A42=subs(simplify(diff(averaged_dvC2_dt,iL2)),
    [iL1 iL2 vC1 vC2 d io],[IL1 IL2 VC1 VC2 D IO]);
A43=subs(simplify(diff(averaged_dvC2_dt,vC1)),
    [iL1 iL2 vC1 vC2 d io],[IL1 IL2 VC1 VC2 D IO]);
A44=subs(simplify(diff(averaged_dvC2_dt,vC2)),
    [iL1 iL2 vC1 vC2 d io],[IL1 IL2 VC1 VC2 D IO]);

AA=eval([A11 A12 A13 A14;
         A21 A22 A23 A24;
         A31 A32 A33 A34;
         A41 A42 A43 A44]);

%Calculating the matrix B
B11=subs(simplify(diff(averaged_diL1_dt,io)),
    [iL1 iL2 vC1 vC2 d vD io vg],[IL1 IL2 VC1 VC2 D VD IO VG]);
B12=subs(simplify(diff(averaged_diL1_dt,vg)),
    [iL1 iL2 vC1 vC2 d vD io vg],[IL1 IL2 VC1 VC2 D VD IO VG]);
B13=subs(simplify(diff(averaged_diL1_dt,d)),
    [iL1 iL2 vC1 vC2 d vD io vg],[IL1 IL2 VC1 VC2 D VD IO VG]);

B21=subs(simplify(diff(averaged_diL2_dt,io)),
    [iL1 iL2 vC1 vC2 d vD io vg],[IL1 IL2 VC1 VC2 D VD IO VG]);
B22=subs(simplify(diff(averaged_diL2_dt,vg)),
    [iL1 iL2 vC1 vC2 d vD io vg],[IL1 IL2 VC1 VC2 D VD IO VG]);
```

```
B23=subs(simplify(diff(averaged_diL2_dt,d)),
    [iL1 iL2 vC1 vC2 d vD io vg],[IL1 IL2 VC1 VC2 D VD IO VG]);

B31=subs(simplify(diff(averaged_dvC1_dt,io)),
    [iL1 iL2 vC1 vC2 d vD io vg],[IL1 IL2 VC1 VC2 D VD IO VG]);
B32=subs(simplify(diff(averaged_dvC1_dt,vg)),
    [iL1 iL2 vC1 vC2 d vD io vg],[IL1 IL2 VC1 VC2 D VD IO VG]);
B33=subs(simplify(diff(averaged_dvC1_dt,d)),
    [iL1 iL2 vC1 vC2 d vD io vg],[IL1 IL2 VC1 VC2 D VD IO VG]);

B41=subs(simplify(diff(averaged_dvC2_dt,io)),
    [iL1 iL2 vC1 vC2 d vD io vg],[IL1 IL2 VC1 VC2 D VD IO VG]);
B42=subs(simplify(diff(averaged_dvC2_dt,vg)),
    [iL1 iL2 vC1 vC2 d vD io vg],[IL1 IL2 VC1 VC2 D VD IO VG]);
B43=subs(simplify(diff(averaged_dvC2_dt,d)),
    [iL1 iL2 vC1 vC2 d vD io vg],[IL1 IL2 VC1 VC2 D VD IO VG]);

BB=eval([B11 B12 B13;
         B21 B22 B23;
         B31 B32 B33;
         B41 B42 B43]);

%Calculating the matrix C
C11=subs(simplify(diff(averaged_vo,iL1)),[iL1 iL2 vC1 vC2 d io],
    [IL1 IL2 VC1 VC2 D IO]);
C12=subs(simplify(diff(averaged_vo,iL2)),[iL1 iL2 vC1 vC2 d io],
    [IL1 IL2 VC1 VC2 D IO]);
C13=subs(simplify(diff(averaged_vo,vC1)),[iL1 iL2 vC1 vC2 d io],
    [IL1 IL2 VC1 VC2 D IO]);
C14=subs(simplify(diff(averaged_vo,vC2)),[iL1 iL2 vC1 vC2 d io],
    [IL1 IL2 VC1 VC2 D IO]);

CC=eval([C11 C12 C13 C14]);

D11=subs(simplify(diff(averaged_vo,io)),
    [iL1 iL2 vC1 vC2 d vD io vg],[IL1 IL2 VC1 VC2 D VD IO VG]);
D12=subs(simplify(diff(averaged_vo,vg)),
    [iL1 iL2 vC1 vC2 d vD io vg],[IL1 IL2 VC1 VC2 D VD IO VG]);
D13=subs(simplify(diff(averaged_vo,d)),
```

```
    [iL1 iL2 vC1 vC2 d vD io vg],[IL1 IL2 VC1 VC2 D VD IO VG]);

%Calculating the matrix D
DD=eval([D11 D12 D13]);

%Producing the State Space Model and obtaining the small
%signal transfer functions
sys=ss(AA,BB,CC,DD);
sys.inputname={'io';'vg';'d'};
sys.outputname={'vo'};

vo_io=tf(sys(1,1)); %Output impedance transfer function
                    %vo(s)/io(s)
vo_vg=tf(sys(1,2)); %vo(s)/vg(s)
vo_d=tf(sys(1,3));  %Control-to-output(vo(s)/d(s))

if n==1
    %there is no variable names "array" in the first
    %running of the loop.
    %variable "array" is initialize in the first running of loop.
    array=vo_d;

else
    array=stack(1,array,vo_d);
end

end

omega=logspace(-1,5,200);
array_frd=frd(array,omega);
relerr = (vo_d_nominal-array_frd)/vo_d_nominal;
[P,Info]=ucover(array_frd,vo_d_nominal,2);
    %fits a second order transfer function.
bodemag(relerr,'b--',Info.W1,'r',{0.1,100000});
```

Obtained results are shown in Figs. 2.34 and 2.35.

Figure 2.36 compares the results obtained using the manual method and the result obtained using the "ucover" command.

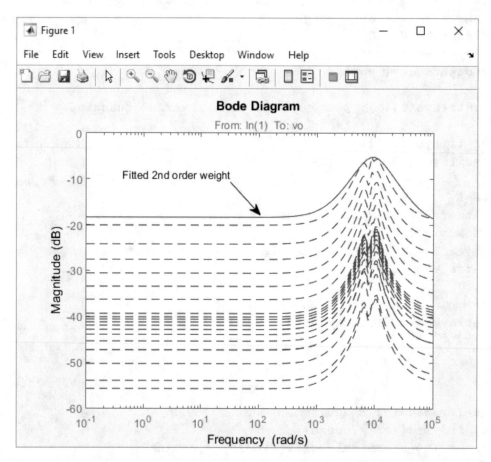

Figure 2.34: Obtaining the multiplicative uncertainty weight for $\frac{v_o(s)}{d(s)}$ with the aid of "ucover" command.

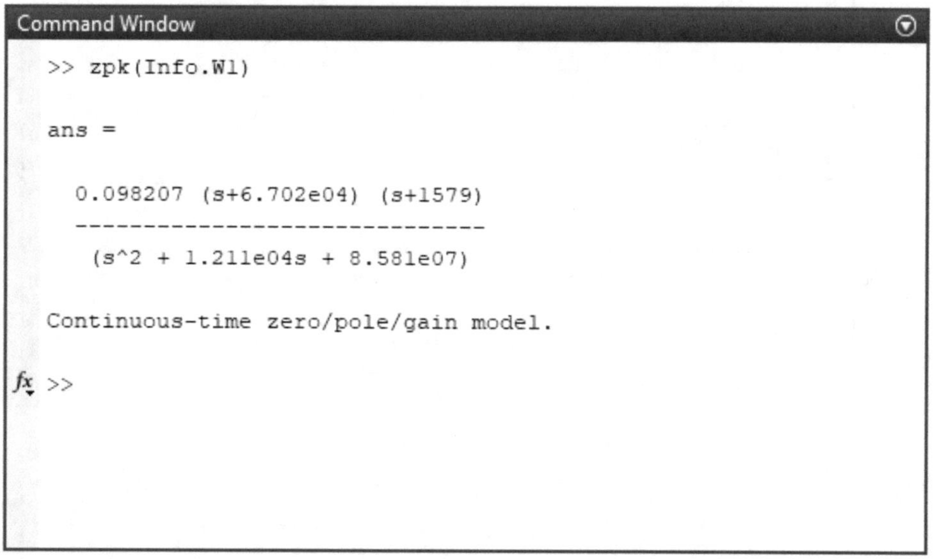

```
Command Window                                                    ⊙

   >> zpk(Info.W1)

   ans =

     0.098207 (s+6.702e04) (s+1579)
     ------------------------------
       (s^2 + 1.211e04s + 8.581e07)

   Continuous-time zero/pole/gain model.

fx >>
```

Figure 2.35: Equation of obtained multiplicative uncertainty weight.

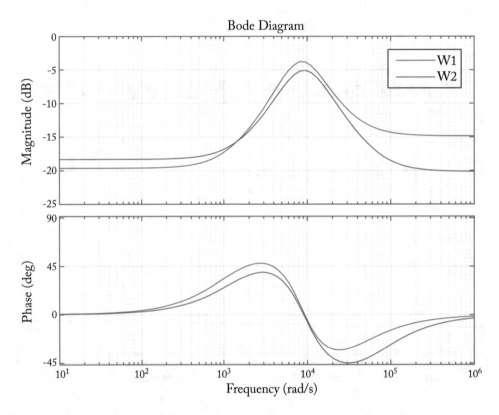

Figure 2.36: W1 is the multiplicative uncertainty weight produced using the manual method. W2 is the multiplicative uncertainty weight produced using the "ucover" command.

Multiplicative uncertainty weights for $\frac{v_o(s)}{v_g(s)}$ and $\frac{v_o(s)}{i_o(s)}$ transfer functions can be obtained in the same way. Following code uses the "ucover" command to extract the multiplicative uncertainty weights for the $\frac{v_o(s)}{v_g(s)}$ and $\frac{v_o(s)}{i_o(s)}$ tranfer functions.

```
%This program study the effect of change in load
%load changes as follows:
%
%   Rnominal=6 ohm
% 1 ohm < R < 11 ohm
%
%we want to obtain the multiplicative
%uncertainty representation of the vo(s)/d(s),
%vo(s)/vg(s) and vo(s)/io(s)
%We use the "ucover" command here.

clc
clear all
close all

%Desired load range for sweep
Rnominal=6;
Rmin=1;
R_delta=.5;
Rmax=11;

VG=20;          %Value of input DC source
rg=0;           %Internal resistance of input DC source
rds=.01;        %MOSFET on resistance
C1=100e-6;      %Capacitor C1 value
C2=220e-6;      %Capacitor C2 value
rC1=.19;        %Capacitor C1 Equivalent Series Resistance(ESR)
rC2=.095;       %Capacitor C2 Equivalent Series Resistance(ESR)
L1=100e-6;      %Inductor L1 value
L2=55e-6;       %Inductor L2 value
rL1=1e-3;       %Inductor L1 Equivalent Series Resistance(ESR)
rL2=.55e-3;     %Inductor L2 Equivalent Series Resistance(ESR)
rD=.01;         %Diode series resistance
VD=.7;          %Diode voltage drop
D=.23;          %Duty cylce
```

```
IO=0;        %Average value of output current source
fsw=100e3;   %Switching frequency

%Nominal transfer functions for R=Rnominal=6 ohm
%These results are obtained in the previous analysis.
s=tf('s');
DEN=(s^2+2239*s+4.76e7)*(s^2+2767*s+1.026e8);
vo_io_nominal=-.093519*(s+4.785e4)*(s+1163)*
   (s^2+1396*s+6.882e7)/DEN;
vo_vg_nominal=391.08*(s+4.785e4)*(s^2+1473*s+7.7e7)/DEN;
vo_d_nominal=43775*(s+4.785e4)*(s^2+1371*s+7.696e7)/DEN;

n=0;
N=length([Rmin:R_delta:Rmax]);

for R=[Rmin:R_delta:Rmax]

if R==Rnominal
    continue
end

n=n+1;
disp('Percentage of work done:')
disp(n/N*100) %shows the progress of the loop

syms iL1 iL2 vC1 vC2 io vg vD d
% iL1: Inductor L1 current
% iL2: Inductor L2 current
% vC1: Capacitor C1 voltage
% vC2: Capacitor C2 voltage
% io : Output current source
% vg : Input DC source
% vD : Diode voltage drop
% d  : Duty cycle

%Closed MOSFET Equations
diL1_dt_MOSFET_close=(-(rL1+rg+rds)*iL1-(rg+rds)*iL2+vg)/L1;
diL2_dt_MOSFET_close=(-(rg+rds)*iL1-(rg+rds+rC1+rL2+R*rC2/
   (R+rC2))*iL2+vC1-R/(R+rC2)*vC2+R*rC2/(R+rC2)*io+vg)/L2;
```

```
dvC1_dt_MOSFET_close=(-iL2)/C1;
dvC2_dt_MOSFET_close=(R/(R+rC2)*iL2-1/(R+rC2)*vC2-R/
   (R+rC2)*io)/C2;
vo_MOSFET_close=R*rC2/(R+rC2)*iL2+R/(R+rC2)*vC2-R*rC2/(R+rC2)*io;

%Opened MOSFET Equations
diL1_dt_MOSFET_open=(-(rL1+rC1+rD)*iL1-rD*iL2-vC1-vD)/L1;
diL2_dt_MOSFET_open=(-rD*iL1-(rD+rL2+R*rC2/(R+rC2))*iL2-R/
   (R+rC2)*vC2+R*rC2/(R+rC2)*io-vD)/L2;
dvC1_dt_MOSFET_open=(iL1)/C1;
dvC2_dt_MOSFET_open=(R/(R+rC2)*iL2-1/(R+rC2)*vC2-R/
   (R+rC2)*io)/C2;
vo_MOSFET_open=R*rC2/(R+rC2)*iL2+R/(R+rC2)*vC2-R*rC2/(R+rC2)*io;

%Averaging
averaged_diL1_dt=simplify(d*diL1_dt_MOSFET_close+(1-d)*
   diL1_dt_MOSFET_open);
averaged_diL2_dt=simplify(d*diL2_dt_MOSFET_close+(1-d)*
   diL2_dt_MOSFET_open);
averaged_dvC1_dt=simplify(d*dvC1_dt_MOSFET_close+(1-d)*
   dvC1_dt_MOSFET_open);
averaged_dvC2_dt=simplify(d*dvC2_dt_MOSFET_close+(1-d)*
   dvC2_dt_MOSFET_open);
averaged_vo=simplify(d*vo_MOSFET_close+(1-d)*vo_MOSFET_open);

%Substituting the steady values of input DC voltage source,
%Diode voltage drop, Duty cycle and output current source
%and calculating the DC operating point
right_side_of_averaged_diL1_dt=subs(averaged_diL1_dt,
   [vg vD d io],[VG VD D IO]);
right_side_of_averaged_diL2_dt=subs(averaged_diL2_dt,
   [vg vD d io],[VG VD D IO]);
right_side_of_averaged_dvC1_dt=subs(averaged_dvC1_dt,
   [vg vD d io],[VG VD D IO]);
right_side_of_averaged_dvC2_dt=subs(averaged_dvC2_dt,
   [vg vD d io],[VG VD D IO]);

DC_OPERATING_POINT=
solve(right_side_of_averaged_diL1_dt==0,
```

```
    right_side_of_averaged_diL2_dt==0,
    right_side_of_averaged_dvC1_dt==0,
    right_side_of_averaged_dvC2_dt==0,'iL1','iL2','vC1','vC2');

IL1=eval(DC_OPERATING_POINT.iL1);
IL2=eval(DC_OPERATING_POINT.iL2);
VC1=eval(DC_OPERATING_POINT.vC1);
VC2=eval(DC_OPERATING_POINT.vC2);
VO=eval(subs(averaged_vo,[iL1 iL2 vC1 vC2 io],
    [IL1 IL2 VC1 VC2 IO]));

%Linearizing the averaged equations around the DC
%operating point. We want to obtain the matrix A, B, C, and D
%      .
%      x=Ax+Bu
%      y=Cx+Du
%
%where,
%      x=[iL1 iL2 vC1 vC2]'
%      u=[io vg d]'
%Since we used the variables D for steady state duty
%ratio and C to show the capacitors values we use AA,
%BB, CC, and DD instead of A, B, C, and D.

%Calculating the matrix A
A11=subs(simplify(diff(averaged_diL1_dt,iL1)),
    [iL1 iL2 vC1 vC2 d io],[IL1 IL2 VC1 VC2 D IO]);
A12=subs(simplify(diff(averaged_diL1_dt,iL2)),
    [iL1 iL2 vC1 vC2 d io],[IL1 IL2 VC1 VC2 D IO]);
A13=subs(simplify(diff(averaged_diL1_dt,vC1)),
    [iL1 iL2 vC1 vC2 d io],[IL1 IL2 VC1 VC2 D IO]);
A14=subs(simplify(diff(averaged_diL1_dt,vC2)),
    [iL1 iL2 vC1 vC2 d io],[IL1 IL2 VC1 VC2 D IO]);

A21=subs(simplify(diff(averaged_diL2_dt,iL1)),
    [iL1 iL2 vC1 vC2 d io],[IL1 IL2 VC1 VC2 D IO]);
A22=subs(simplify(diff(averaged_diL2_dt,iL2)),
    [iL1 iL2 vC1 vC2 d io],[IL1 IL2 VC1 VC2 D IO]);
A23=subs(simplify(diff(averaged_diL2_dt,vC1)),
```

```
      [iL1 iL2 vC1 vC2 d io],[IL1 IL2 VC1 VC2 D IO]);
A24=subs(simplify(diff(averaged_diL2_dt,vC2)),
      [iL1 iL2 vC1 vC2 d io],[IL1 IL2 VC1 VC2 D IO]);

A31=subs(simplify(diff(averaged_dvC1_dt,iL1)),
      [iL1 iL2 vC1 vC2 d io],[IL1 IL2 VC1 VC2 D IO]);
A32=subs(simplify(diff(averaged_dvC1_dt,iL2)),
      [iL1 iL2 vC1 vC2 d io],[IL1 IL2 VC1 VC2 D IO]);
A33=subs(simplify(diff(averaged_dvC1_dt,vC1)),
      [iL1 iL2 vC1 vC2 d io],[IL1 IL2 VC1 VC2 D IO]);
A34=subs(simplify(diff(averaged_dvC1_dt,vC2)),
      [iL1 iL2 vC1 vC2 d io],[IL1 IL2 VC1 VC2 D IO]);

A41=subs(simplify(diff(averaged_dvC2_dt,iL1)),
      [iL1 iL2 vC1 vC2 d io],[IL1 IL2 VC1 VC2 D IO]);
A42=subs(simplify(diff(averaged_dvC2_dt,iL2)),
      [iL1 iL2 vC1 vC2 d io],[IL1 IL2 VC1 VC2 D IO]);
A43=subs(simplify(diff(averaged_dvC2_dt,vC1)),
      [iL1 iL2 vC1 vC2 d io],[IL1 IL2 VC1 VC2 D IO]);
A44=subs(simplify(diff(averaged_dvC2_dt,vC2)),
      [iL1 iL2 vC1 vC2 d io],[IL1 IL2 VC1 VC2 D IO]);

AA=eval([A11 A12 A13 A14;
         A21 A22 A23 A24;
         A31 A32 A33 A34;
         A41 A42 A43 A44]);

%Calculating the matrix B
B11=subs(simplify(diff(averaged_diL1_dt,io)),
      [iL1 iL2 vC1 vC2 d vD io vg],[IL1 IL2 VC1 VC2 D VD IO VG]);
B12=subs(simplify(diff(averaged_diL1_dt,vg)),
      [iL1 iL2 vC1 vC2 d vD io vg],[IL1 IL2 VC1 VC2 D VD IO VG]);
B13=subs(simplify(diff(averaged_diL1_dt,d)),
      [iL1 iL2 vC1 vC2 d vD io vg],[IL1 IL2 VC1 VC2 D VD IO VG]);

B21=subs(simplify(diff(averaged_diL2_dt,io)),
      [iL1 iL2 vC1 vC2 d vD io vg],[IL1 IL2 VC1 VC2 D VD IO VG]);
B22=subs(simplify(diff(averaged_diL2_dt,vg)),
      [iL1 iL2 vC1 vC2 d vD io vg],[IL1 IL2 VC1 VC2 D VD IO VG]);
```

```
B23=subs(simplify(diff(averaged_diL2_dt,d)),
    [iL1 iL2 vC1 vC2 d vD io vg],[IL1 IL2 VC1 VC2 D VD IO VG]);

B31=subs(simplify(diff(averaged_dvC1_dt,io)),
    [iL1 iL2 vC1 vC2 d vD io vg],[IL1 IL2 VC1 VC2 D VD IO VG]);
B32=subs(simplify(diff(averaged_dvC1_dt,vg)),
    [iL1 iL2 vC1 vC2 d vD io vg],[IL1 IL2 VC1 VC2 D VD IO VG]);
B33=subs(simplify(diff(averaged_dvC1_dt,d)),
    [iL1 iL2 vC1 vC2 d vD io vg],[IL1 IL2 VC1 VC2 D VD IO VG]);

B41=subs(simplify(diff(averaged_dvC2_dt,io)),
    [iL1 iL2 vC1 vC2 d vD io vg],[IL1 IL2 VC1 VC2 D VD IO VG]);
B42=subs(simplify(diff(averaged_dvC2_dt,vg)),
    [iL1 iL2 vC1 vC2 d vD io vg],[IL1 IL2 VC1 VC2 D VD IO VG]);
B43=subs(simplify(diff(averaged_dvC2_dt,d)),
    [iL1 iL2 vC1 vC2 d vD io vg],[IL1 IL2 VC1 VC2 D VD IO VG]);

BB=eval([B11 B12 B13;
         B21 B22 B23;
         B31 B32 B33;
         B41 B42 B43]);

%Calculating the matrix C
C11=subs(simplify(diff(averaged_vo,iL1)),[iL1 iL2 vC1 vC2 d io],
    [IL1 IL2 VC1 VC2 D IO]);
C12=subs(simplify(diff(averaged_vo,iL2)),[iL1 iL2 vC1 vC2 d io],
    [IL1 IL2 VC1 VC2 D IO]);
C13=subs(simplify(diff(averaged_vo,vC1)),[iL1 iL2 vC1 vC2 d io],
    [IL1 IL2 VC1 VC2 D IO]);
C14=subs(simplify(diff(averaged_vo,vC2)),[iL1 iL2 vC1 vC2 d io],
    [IL1 IL2 VC1 VC2 D IO]);

CC=eval([C11 C12 C13 C14]);

D11=subs(simplify(diff(averaged_vo,io)),
    [iL1 iL2 vC1 vC2 d vD io vg],[IL1 IL2 VC1 VC2 D VD IO VG]);
D12=subs(simplify(diff(averaged_vo,vg)),
    [iL1 iL2 vC1 vC2 d vD io vg],[IL1 IL2 VC1 VC2 D VD IO VG]);
D13=subs(simplify(diff(averaged_vo,d)),
```

```
    [iL1 iL2 vC1 vC2 d vD io vg],[IL1 IL2 VC1 VC2 D VD IO VG]);

%Calculating the matrix D
DD=eval([D11 D12 D13]);

%Producing the State Space Model and obtaining the small
%signal transfer functions
sys=ss(AA,BB,CC,DD);
sys.inputname={'io';'vg';'d'};
sys.outputname={'vo'};

vo_io=tf(sys(1,1)); %Output impedance transfer function
                    %vo(s)/io(s)
vo_vg=tf(sys(1,2)); %vo(s)/vg(s)
vo_d=tf(sys(1,3));  %Control-to-output(vo(s)/d(s))

if n==1
    %there is no variable names "array" in the first
    %running of the loop.
    %variable "array" is initialize in the first running of loop.
    array1=vo_d;
    array2=vo_vg;
    array3=vo_io;
else
    array1=stack(1,array1,vo_d);
    array2=stack(1,array2,vo_vg);
    array3=stack(1,array3,vo_io);
end

end

omega=logspace(-1,5,200);
array1_frd=frd(array1,omega);
array2_frd=frd(array2,omega);
array3_frd=frd(array3,omega);

relerr1 = (vo_d_nominal-array1_frd)/vo_d_nominal;
relerr2 = (vo_vg_nominal-array2_frd)/vo_vg_nominal;
relerr3 = (vo_io_nominal-array3_frd)/vo_io_nominal;
```

```
[P1,Info1]=ucover(array1_frd,vo_d_nominal,2);
[P2,Info2]=ucover(array2_frd,vo_vg_nominal,2);
[P3,Info3]=ucover(array3_frd,vo_io_nominal,2);

figure(1)
bodemag(relerr1,'b--',Info1.W1,'r',{0.1,100000});
figure(2)
bodemag(relerr2,'b--',Info2.W1,'r',{0.1,100000});
figure(3)
bodemag(relerr3,'b--',Info3.W1,'r',{0.1,100000});
```

After running the code, results shown in Figs. 2.37–2.40 are obtained.

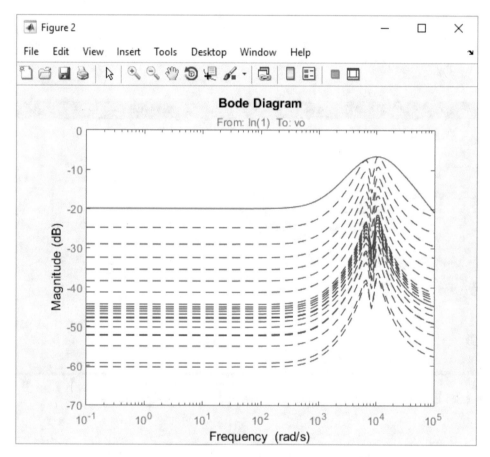

Figure 2.37: Obtaining the multiplicative uncertainty weight for $\frac{v_o(s)}{v_g(s)}$ with the aid of "ucover" command.

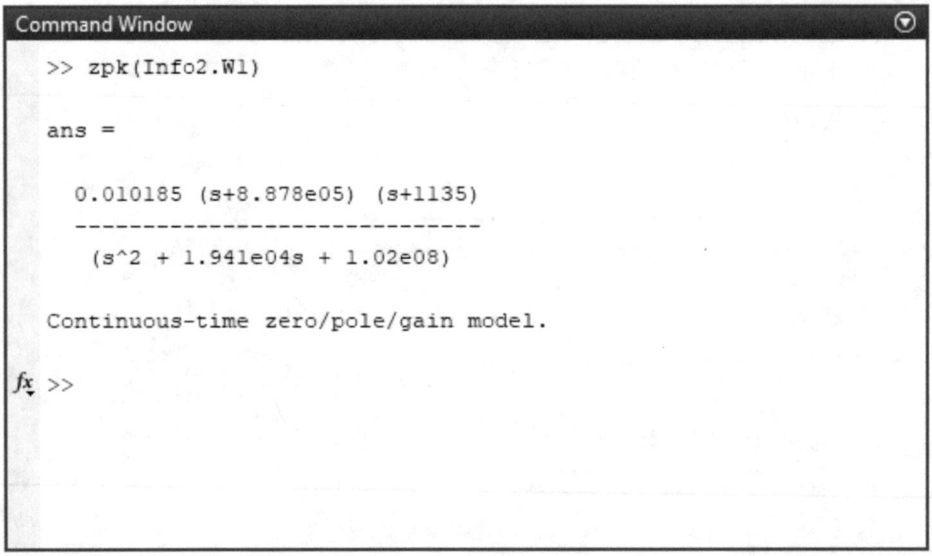

```
Command Window                                              ⊙

   >> zpk(Info2.W1)

   ans =

     0.010185 (s+8.878e05) (s+1135)
     ------------------------------
       (s^2 + 1.941e04s + 1.02e08)

   Continuous-time zero/pole/gain model.

fx >>
```

Figure 2.38: Equation of obtained multiplicative uncertainty weight for $\frac{v_o(s)}{v_g(s)}$.

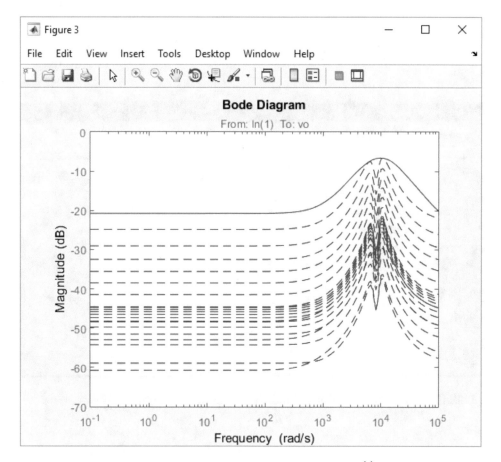

Figure 2.39: Obtaining the multiplicative uncertainty weight for $\frac{v_o(s)}{i_o(s)}$ with the aid of "ucover" command.

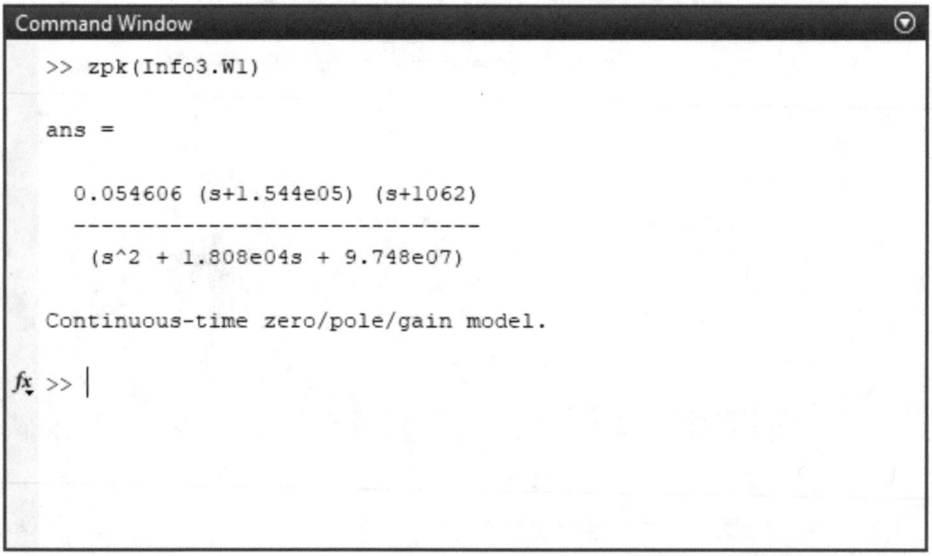

```
Command Window                                                    ⊙

   >> zpk(Info3.W1)

   ans =

     0.054606 (s+1.544e05) (s+1062)
     -------------------------------
      (s^2 + 1.808e04s + 9.748e07)

   Continuous-time zero/pole/gain model.

fx >> |
```

Figure 2.40: Equation of obtained multiplicative uncertainty weight for $\frac{v_o(s)}{i_o(s)}$.

2.9   TESTING THE OBTAINED UNCERTAINTY WEIGHTS

The following code is used to test the obtained weights. The code is composed of two main parts. First part of code, draws converter transfer functions for different loads. The code ask the user to press any key. After pressing a key, the second part of the code starts. The second part uses the commands provided by MATLAB®Robust Control Toolbox. It makes the uncertain model of the converter and draws some of the uncertain object samples on the same plot used by the first part of the code. If the result produced by the second part of the code covers the result procuced by the first part of the code, one can deduce that the model is reliable.

```
%This example shows that the obtained uncertainity
%for vo(s)/d(s) really cover the changes in the vo(s)/d(s).
clc
clear all
close all

%Desired load range for sweep
Rnominal=6;
Rmin=1;
R_delta=.5;
Rmax=11;

VG=20;        %Value of input DC source
rg=0;         %Internal resistance of input DC source
rds=.01;      %MOSFET on resistance
C1=100e-6;    %Capacitor C1 value
C2=220e-6;    %Capacitor C2 value
rC1=.19;      %Capacitor C1 Equivalent Series Resistance(ESR)
rC2=.095;     %Capacitor C2 Equivalent Series Resistance(ESR)
L1=100e-6;    %Inductor L1 value
L2=55e-6;     %Inductor L2 value
rL1=1e-3;     %Inductor L1 Equivalent Series Resistance(ESR)
rL2=.55e-3;   %Inductor L2 Equivalent Series Resistance(ESR)
rD=.01;       %Diode series resistance
VD=.7;        %Diode voltage drop
D=.23;        %Duty cylcle
IO=0;         %Average value of output current source
fsw=100e3;    %Switching frequency

%Nominal transfer functions for R=Rnominal=6 ohm
```

```
%These results are obtained in the previous analysis.
s=tf('s');
DEN=(s^2+2239*s+4.76e7)*(s^2+2767*s+1.026e8);
vo_io_nominal=-.093519*(s+4.785e4)*(s+1163)*
    (s^2+1396*s+6.882e7)/DEN;
vo_vg_nominal=391.08*(s+4.785e4)*(s^2+1473*s+7.7e7)/DEN;
vo_d_nominal=43775*(s+4.785e4)*(s^2+1371*s+7.696e7)/DEN;

n=0;
N=length([Rmin:R_delta:Rmax]);

for R=[Rmin:R_delta:Rmax]

if R==Rnominal
    continue
end

n=n+1;
disp('Percentage of work done:')
disp(n/N*100) %shows the progress of the loop

syms iL1 iL2 vC1 vC2 io vg vD d
% iL1: Inductor L1 current
% iL2: Inductor L2 current
% vC1: Capacitor C1 voltage
% vC2: Capacitor C2 voltage
% io : Output current source
% vg : Input DC source
% vD : Diode voltage drop
% d  : Duty cycle

%Closed MOSFET Equations
diL1_dt_MOSFET_close=(-(rL1+rg+rds)*iL1-(rg+rds)*iL2+vg)/L1;
diL2_dt_MOSFET_close=(-(rg+rds)*iL1-(rg+rds+rC1+rL2+R*rC2/
    (R+rC2))*iL2+vC1-R/(R+rC2)*vC2+R*rC2/(R+rC2)*io+vg)/L2;
dvC1_dt_MOSFET_close=(-iL2)/C1;
dvC2_dt_MOSFET_close=(R/(R+rC2)*iL2-1/(R+rC2)*vC2-R/
    (R+rC2)*io)/C2;
vo_MOSFET_close=R*rC2/(R+rC2)*iL2+R/(R+rC2)*vC2-R*rC2/(R+rC2)*io;
```

```
%Opened MOSFET Equations
diL1_dt_MOSFET_open=(-(rL1+rC1+rD)*iL1-rD*iL2-vC1-vD)/L1;
diL2_dt_MOSFET_open=(-rD*iL1-(rD+rL2+R*rC2/(R+rC2))*iL2-R/
   (R+rC2)*vC2+R*rC2/(R+rC2)*io-vD)/L2;
dvC1_dt_MOSFET_open=(iL1)/C1;
dvC2_dt_MOSFET_open=(R/(R+rC2)*iL2-1/(R+rC2)*vC2-R/
   (R+rC2)*io)/C2;
vo_MOSFET_open=R*rC2/(R+rC2)*iL2+R/(R+rC2)*vC2-R*rC2/(R+rC2)*io;

%Averaging
averaged_diL1_dt=simplify(d*diL1_dt_MOSFET_close+(1-d)*
   diL1_dt_MOSFET_open);
averaged_diL2_dt=simplify(d*diL2_dt_MOSFET_close+(1-d)*
   diL2_dt_MOSFET_open);
averaged_dvC1_dt=simplify(d*dvC1_dt_MOSFET_close+(1-d)*
   dvC1_dt_MOSFET_open);
averaged_dvC2_dt=simplify(d*dvC2_dt_MOSFET_close+(1-d)*
   dvC2_dt_MOSFET_open);
averaged_vo=simplify(d*vo_MOSFET_close+(1-d)*vo_MOSFET_open);

%Substituting the steady values of input DC voltage source,
%Diode voltage drop, Duty cycle and output current source
%and calculating the DC operating point
right_side_of_averaged_diL1_dt=subs(averaged_diL1_dt,
   [vg vD d io],[VG VD D IO]);
right_side_of_averaged_diL2_dt=subs(averaged_diL2_dt,
   [vg vD d io],[VG VD D IO]);
right_side_of_averaged_dvC1_dt=subs(averaged_dvC1_dt,
   [vg vD d io],[VG VD D IO]);
right_side_of_averaged_dvC2_dt=subs(averaged_dvC2_dt,
   [vg vD d io],[VG VD D IO]);

DC_OPERATING_POINT=
solve(right_side_of_averaged_diL1_dt==0,
   right_side_of_averaged_diL2_dt==0,
   right_side_of_averaged_dvC1_dt==0,
   right_side_of_averaged_dvC2_dt==0,'iL1','iL2','vC1','vC2');
```

```
IL1=eval(DC_OPERATING_POINT.iL1);
IL2=eval(DC_OPERATING_POINT.iL2);
VC1=eval(DC_OPERATING_POINT.vC1);
VC2=eval(DC_OPERATING_POINT.vC2);
VO=eval(subs(averaged_vo,[iL1 iL2 vC1 vC2 io],
    [IL1 IL2 VC1 VC2 IO]));

%Linearizing the averaged equations around the DC
%operating point. We want to obtain the matrix A, B, C, and D
%       .
%       x=Ax+Bu
%       y=Cx+Du
%
%where,
%       x=[iL1 iL2 vC1 vC2]'
%       u=[io vg d]'
%Since we used the variables D for steady state duty
%ratio and C to show the capacitors values we use AA,
%BB, CC, and DD instead of A, B, C, and D.

%Calculating the matrix A
A11=subs(simplify(diff(averaged_diL1_dt,iL1)),
    [iL1 iL2 vC1 vC2 d io],[IL1 IL2 VC1 VC2 D IO]);
A12=subs(simplify(diff(averaged_diL1_dt,iL2)),
    [iL1 iL2 vC1 vC2 d io],[IL1 IL2 VC1 VC2 D IO]);
A13=subs(simplify(diff(averaged_diL1_dt,vC1)),
    [iL1 iL2 vC1 vC2 d io],[IL1 IL2 VC1 VC2 D IO]);
A14=subs(simplify(diff(averaged_diL1_dt,vC2)),
    [iL1 iL2 vC1 vC2 d io],[IL1 IL2 VC1 VC2 D IO]);

A21=subs(simplify(diff(averaged_diL2_dt,iL1)),
    [iL1 iL2 vC1 vC2 d io],[IL1 IL2 VC1 VC2 D IO]);
A22=subs(simplify(diff(averaged_diL2_dt,iL2)),
    [iL1 iL2 vC1 vC2 d io],[IL1 IL2 VC1 VC2 D IO]);
A23=subs(simplify(diff(averaged_diL2_dt,vC1)),
    [iL1 iL2 vC1 vC2 d io],[IL1 IL2 VC1 VC2 D IO]);
A24=subs(simplify(diff(averaged_diL2_dt,vC2)),
    [iL1 iL2 vC1 vC2 d io],[IL1 IL2 VC1 VC2 D IO]);
```

```
A31=subs(simplify(diff(averaged_dvC1_dt,iL1)),
    [iL1 iL2 vC1 vC2 d io],[IL1 IL2 VC1 VC2 D IO]);
A32=subs(simplify(diff(averaged_dvC1_dt,iL2)),
    [iL1 iL2 vC1 vC2 d io],[IL1 IL2 VC1 VC2 D IO]);
A33=subs(simplify(diff(averaged_dvC1_dt,vC1)),
    [iL1 iL2 vC1 vC2 d io],[IL1 IL2 VC1 VC2 D IO]);
A34=subs(simplify(diff(averaged_dvC1_dt,vC2)),
    [iL1 iL2 vC1 vC2 d io],[IL1 IL2 VC1 VC2 D IO]);

A41=subs(simplify(diff(averaged_dvC2_dt,iL1)),
    [iL1 iL2 vC1 vC2 d io],[IL1 IL2 VC1 VC2 D IO]);
A42=subs(simplify(diff(averaged_dvC2_dt,iL2)),
    [iL1 iL2 vC1 vC2 d io],[IL1 IL2 VC1 VC2 D IO]);
A43=subs(simplify(diff(averaged_dvC2_dt,vC1)),
    [iL1 iL2 vC1 vC2 d io],[IL1 IL2 VC1 VC2 D IO]);
A44=subs(simplify(diff(averaged_dvC2_dt,vC2)),
    [iL1 iL2 vC1 vC2 d io],[IL1 IL2 VC1 VC2 D IO]);

AA=eval([A11 A12 A13 A14;
         A21 A22 A23 A24;
         A31 A32 A33 A34;
         A41 A42 A43 A44]);

%Calculating the matrix B
B11=subs(simplify(diff(averaged_diL1_dt,io)),
    [iL1 iL2 vC1 vC2 d vD io vg],[IL1 IL2 VC1 VC2 D VD IO VG]);
B12=subs(simplify(diff(averaged_diL1_dt,vg)),
    [iL1 iL2 vC1 vC2 d vD io vg],[IL1 IL2 VC1 VC2 D VD IO VG]);
B13=subs(simplify(diff(averaged_diL1_dt,d)),
    [iL1 iL2 vC1 vC2 d vD io vg],[IL1 IL2 VC1 VC2 D VD IO VG]);

B21=subs(simplify(diff(averaged_diL2_dt,io)),
    [iL1 iL2 vC1 vC2 d vD io vg],[IL1 IL2 VC1 VC2 D VD IO VG]);
B22=subs(simplify(diff(averaged_diL2_dt,vg)),
    [iL1 iL2 vC1 vC2 d vD io vg],[IL1 IL2 VC1 VC2 D VD IO VG]);
B23=subs(simplify(diff(averaged_diL2_dt,d)),
    [iL1 iL2 vC1 vC2 d vD io vg],[IL1 IL2 VC1 VC2 D VD IO VG]);

B31=subs(simplify(diff(averaged_dvC1_dt,io)),
```

```
    [iL1 iL2 vC1 vC2 d vD io vg],[IL1 IL2 VC1 VC2 D VD IO VG]);
B32=subs(simplify(diff(averaged_dvC1_dt,vg)),
    [iL1 iL2 vC1 vC2 d vD io vg],[IL1 IL2 VC1 VC2 D VD IO VG]);
B33=subs(simplify(diff(averaged_dvC1_dt,d)),
    [iL1 iL2 vC1 vC2 d vD io vg],[IL1 IL2 VC1 VC2 D VD IO VG]);

B41=subs(simplify(diff(averaged_dvC2_dt,io)),
    [iL1 iL2 vC1 vC2 d vD io vg],[IL1 IL2 VC1 VC2 D VD IO VG]);
B42=subs(simplify(diff(averaged_dvC2_dt,vg)),
    [iL1 iL2 vC1 vC2 d vD io vg],[IL1 IL2 VC1 VC2 D VD IO VG]);
B43=subs(simplify(diff(averaged_dvC2_dt,d)),
    [iL1 iL2 vC1 vC2 d vD io vg],[IL1 IL2 VC1 VC2 D VD IO VG]);

BB=eval([B11 B12 B13;
         B21 B22 B23;
         B31 B32 B33;
         B41 B42 B43]);

%Calculating the matrix C
C11=subs(simplify(diff(averaged_vo,iL1)),[iL1 iL2 vC1 vC2 d io],
    [IL1 IL2 VC1 VC2 D IO]);
C12=subs(simplify(diff(averaged_vo,iL2)),[iL1 iL2 vC1 vC2 d io],
    [IL1 IL2 VC1 VC2 D IO]);
C13=subs(simplify(diff(averaged_vo,vC1)),[iL1 iL2 vC1 vC2 d io],
    [IL1 IL2 VC1 VC2 D IO]);
C14=subs(simplify(diff(averaged_vo,vC2)),[iL1 iL2 vC1 vC2 d io],
    [IL1 IL2 VC1 VC2 D IO]);

CC=eval([C11 C12 C13 C14]);

D11=subs(simplify(diff(averaged_vo,io)),
    [iL1 iL2 vC1 vC2 d vD io vg],[IL1 IL2 VC1 VC2 D VD IO VG]);
D12=subs(simplify(diff(averaged_vo,vg)),
    [iL1 iL2 vC1 vC2 d vD io vg],[IL1 IL2 VC1 VC2 D VD IO VG]);
D13=subs(simplify(diff(averaged_vo,d)),
    [iL1 iL2 vC1 vC2 d vD io vg],[IL1 IL2 VC1 VC2 D VD IO VG]);

%Calculating the matrix D
DD=eval([D11 D12 D13]);
```

```
%Producing the State Space Model and obtaining the
%small signal transfer functions
sys=ss(AA,BB,CC,DD);
sys.inputname={'io';'vg';'d'};
sys.outputname={'vo'};

vo_io=tf(sys(1,1)); %Output impedance transfer function
                    %vo(s)/io(s)
vo_vg=tf(sys(1,2)); %vo(s)/vg(s)
vo_d=tf(sys(1,3));  %Control-to-output(vo(s)/d(s))

bode(vo_d,{100,1e5}), hold on
end

pause %press any key to continue
disp('Press any key to continue...')

W_vo_d=.098207*(s+6.702e4)*(s+1579)/(s^2+1.211e4*s+8.581e7);
   %The weight obtained in previous sections for vo(s)/d(s)
Delta=ultidyn('Delta',[1 1]);
vo_d_unc=vo_d_nominal*(1+W_vo_d*Delta);
bode(vo_d_unc,'r--')
```

The first part of the code produces the graphic shown in Fig. 2.41. After showing the figure, the user is prompted to press any key.

After pressing a key, the graphic shown in Fig. 2.42 will appear. As shown, the uncertain model covers the plot produced by the first part of the code. So, one can deduce that the formed model is reliable.

Reliability of uncertain models obtained for $\frac{v_o(s)}{v_g(s)}$ and $\frac{v_o(s)}{i_o(s)}$ can be tested in the same way. The following code is the modified version of the previous code. This modified code tests the reliability of the uncertain models obtained for the $\frac{v_o(s)}{v_g(s)}$ and $\frac{v_o(s)}{i_o(s)}$ as well.

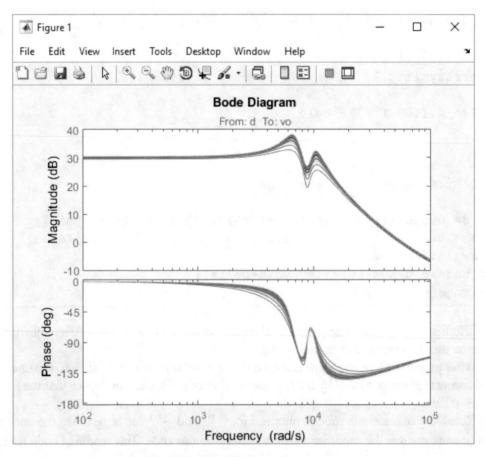

Figure 2.41: Effect of load changes on the $\frac{v_o(s)}{d(s)}$ transfer function.

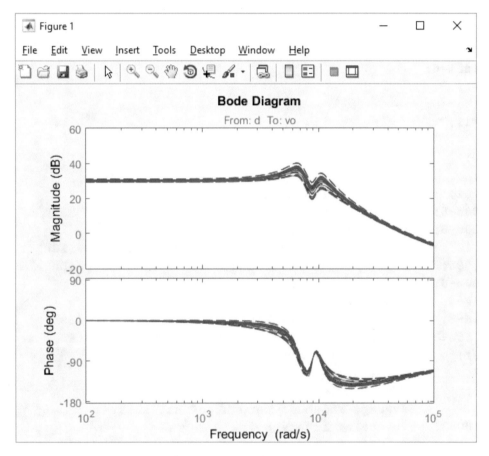

Figure 2.42: The random transfer functions produced according to the developed model cover the transfer functions shown in Fig. 2.41.

```
%This example shows that the obtained uncertainity really
%cover the changes in the transfer functions
%vo(s)/d(s), vo(s)/vg(s) and vo(s)/io(s) are studied here.
clc
clear all
close all

%Desired load range for sweep
Rnominal=6;
Rmin=1;
R_delta=.5;
Rmax=11;

VG=20;        %Value of input DC source
rg=0;         %Internal resistance of input DC source
rds=.01;      %MOSFET on resistance
C1=100e-6;    %Capacitor C1 value
C2=220e-6;    %Capacitor C2 value
rC1=.19;      %Capacitor C1 Equivalent Series Resistance(ESR)
rC2=.095;     %Capacitor C2 Equivalent Series Resistance(ESR)
L1=100e-6;    %Inductor L1 value
L2=55e-6;     %Inductor L2 value
rL1=1e-3;     %Inductor L1 Equivalent Series Resistance(ESR)
rL2=.55e-3;   %Inductor L2 Equivalent Series Resistance(ESR)
rD=.01;       %Diode series resistance
VD=.7;        %Diode voltage drop
D=.23;        %Duty cylcle
IO=0;         %Average value of output current source
fsw=100e3;    %Switching frequency

%Nominal transfer functions for R=Rnominal=6 ohm
%These results are obtained in the previous analysis.
s=tf('s');
DEN=(s^2+2239*s+4.76e7)*(s^2+2767*s+1.026e8);
vo_io_nominal=-.093519*(s+4.785e4)*(s+1163)*
   (s^2+1396*s+6.882e7)/DEN;
vo_vg_nominal=391.08*(s+4.785e4)*(s^2+1473*s+7.7e7)/DEN;
vo_d_nominal=43775*(s+4.785e4)*(s^2+1371*s+7.696e7)/DEN;
```

```
n=0;
N=length([Rmin:R_delta:Rmax]);

for R=[Rmin:R_delta:Rmax]

if R==Rnominal
    continue
end

n=n+1;
disp('Percentage of work done:')
disp(n/N*100) %shows the progress of the loop

syms iL1 iL2 vC1 vC2 io vg vD d
% iL1: Inductor L1 current
% iL2: Inductor L2 current
% vC1: Capacitor C1 voltage
% vC2: Capacitor C2 voltage
% io : Output current source
% vg : Input DC source
% vD : Diode voltage drop
% d  : Duty cycle

%Closed MOSFET Equations
diL1_dt_MOSFET_close=(-(rL1+rg+rds)*iL1-(rg+rds)*iL2+vg)/L1;
diL2_dt_MOSFET_close=(-(rg+rds)*iL1-(rg+rds+rC1+rL2+R*rC2/
    (R+rC2))*iL2+vC1-R/(R+rC2)*vC2+R*rC2/(R+rC2)*io+vg)/L2;
dvC1_dt_MOSFET_close=(-iL2)/C1;
dvC2_dt_MOSFET_close=(R/(R+rC2)*iL2-1/(R+rC2)*vC2-R/
    (R+rC2)*io)/C2;
vo_MOSFET_close=R*rC2/(R+rC2)*iL2+R/(R+rC2)*vC2-R*rC2/(R+rC2)*io;

%Opened MOSFET Equations
diL1_dt_MOSFET_open=(-(rL1+rC1+rD)*iL1-rD*iL2-vC1-vD)/L1;
diL2_dt_MOSFET_open=(-rD*iL1-(rD+rL2+R*rC2/(R+rC2))*iL2-R/
    (R+rC2)*vC2+R*rC2/(R+rC2)*io-vD)/L2;
dvC1_dt_MOSFET_open=(iL1)/C1;
dvC2_dt_MOSFET_open=(R/(R+rC2)*iL2-1/(R+rC2)*vC2-R/
```

```
    (R+rC2)*io)/C2;
vo_MOSFET_open=R*rC2/(R+rC2)*iL2+R/(R+rC2)*vC2-R*rC2/(R+rC2)*io;

%Averaging
averaged_diL1_dt=simplify(d*diL1_dt_MOSFET_close+(1-d)*
    diL1_dt_MOSFET_open);
averaged_diL2_dt=simplify(d*diL2_dt_MOSFET_close+(1-d)*
    diL2_dt_MOSFET_open);
averaged_dvC1_dt=simplify(d*dvC1_dt_MOSFET_close+(1-d)*
    dvC1_dt_MOSFET_open);
averaged_dvC2_dt=simplify(d*dvC2_dt_MOSFET_close+(1-d)*
    dvC2_dt_MOSFET_open);
averaged_vo=simplify(d*vo_MOSFET_close+(1-d)*vo_MOSFET_open);

%Substituting the steady values of input DC voltage source,
%Diode voltage drop, Duty cycle and output current source
%and calculating the DC operating point
right_side_of_averaged_diL1_dt=subs(averaged_diL1_dt,
    [vg vD d io],[VG VD D IO]);
right_side_of_averaged_diL2_dt=subs(averaged_diL2_dt,
    [vg vD d io],[VG VD D IO]);
right_side_of_averaged_dvC1_dt=subs(averaged_dvC1_dt,
    [vg vD d io],[VG VD D IO]);
right_side_of_averaged_dvC2_dt=subs(averaged_dvC2_dt,
    [vg vD d io],[VG VD D IO]);

DC_OPERATING_POINT=
solve(right_side_of_averaged_diL1_dt==0,
    right_side_of_averaged_diL2_dt==0,
    right_side_of_averaged_dvC1_dt==0,
    right_side_of_averaged_dvC2_dt==0,'iL1','iL2','vC1','vC2');

IL1=eval(DC_OPERATING_POINT.iL1);
IL2=eval(DC_OPERATING_POINT.iL2);
VC1=eval(DC_OPERATING_POINT.vC1);
VC2=eval(DC_OPERATING_POINT.vC2);
VO=eval(subs(averaged_vo,[iL1 iL2 vC1 vC2 io],
    [IL1 IL2 VC1 VC2 IO]));
```

```
%Linearizing the averaged equations around the DC
%operating point. We want to obtain the matrix A, B, C, and D
%       .
%      x=Ax+Bu
%      y=Cx+Du
%
%where,
%      x=[iL1 iL2 vC1 vC2]'
%      u=[io vg d]'
%Since we used the variables D for steady state duty ratio
%and C to show the capacitors values we use AA, BB, CC and
%DD instead of A, B, C, and D.

%Calculating the matrix A
A11=subs(simplify(diff(averaged_diL1_dt,iL1)),
    [iL1 iL2 vC1 vC2 d io],[IL1 IL2 VC1 VC2 D IO]);
A12=subs(simplify(diff(averaged_diL1_dt,iL2)),
    [iL1 iL2 vC1 vC2 d io],[IL1 IL2 VC1 VC2 D IO]);
A13=subs(simplify(diff(averaged_diL1_dt,vC1)),
    [iL1 iL2 vC1 vC2 d io],[IL1 IL2 VC1 VC2 D IO]);
A14=subs(simplify(diff(averaged_diL1_dt,vC2)),
    [iL1 iL2 vC1 vC2 d io],[IL1 IL2 VC1 VC2 D IO]);

A21=subs(simplify(diff(averaged_diL2_dt,iL1)),
    [iL1 iL2 vC1 vC2 d io],[IL1 IL2 VC1 VC2 D IO]);
A22=subs(simplify(diff(averaged_diL2_dt,iL2)),
    [iL1 iL2 vC1 vC2 d io],[IL1 IL2 VC1 VC2 D IO]);
A23=subs(simplify(diff(averaged_diL2_dt,vC1)),
    [iL1 iL2 vC1 vC2 d io],[IL1 IL2 VC1 VC2 D IO]);
A24=subs(simplify(diff(averaged_diL2_dt,vC2)),
    [iL1 iL2 vC1 vC2 d io],[IL1 IL2 VC1 VC2 D IO]);

A31=subs(simplify(diff(averaged_dvC1_dt,iL1)),
    [iL1 iL2 vC1 vC2 d io],[IL1 IL2 VC1 VC2 D IO]);
A32=subs(simplify(diff(averaged_dvC1_dt,iL2)),
    [iL1 iL2 vC1 vC2 d io],[IL1 IL2 VC1 VC2 D IO]);
A33=subs(simplify(diff(averaged_dvC1_dt,vC1)),
    [iL1 iL2 vC1 vC2 d io],[IL1 IL2 VC1 VC2 D IO]);
A34=subs(simplify(diff(averaged_dvC1_dt,vC2)),
```

```
        [iL1 iL2 vC1 vC2 d io],[IL1 IL2 VC1 VC2 D IO]);

A41=subs(simplify(diff(averaged_dvC2_dt,iL1)),
    [iL1 iL2 vC1 vC2 d io],[IL1 IL2 VC1 VC2 D IO]);
A42=subs(simplify(diff(averaged_dvC2_dt,iL2)),
    [iL1 iL2 vC1 vC2 d io],[IL1 IL2 VC1 VC2 D IO]);
A43=subs(simplify(diff(averaged_dvC2_dt,vC1)),
    [iL1 iL2 vC1 vC2 d io],[IL1 IL2 VC1 VC2 D IO]);
A44=subs(simplify(diff(averaged_dvC2_dt,vC2)),
    [iL1 iL2 vC1 vC2 d io],[IL1 IL2 VC1 VC2 D IO]);


AA=eval([A11 A12 A13 A14;
         A21 A22 A23 A24;
         A31 A32 A33 A34;
         A41 A42 A43 A44]);

%Calculating the matrix B
B11=subs(simplify(diff(averaged_diL1_dt,io)),
    [iL1 iL2 vC1 vC2 d vD io vg],[IL1 IL2 VC1 VC2 D VD IO VG]);
B12=subs(simplify(diff(averaged_diL1_dt,vg)),
    [iL1 iL2 vC1 vC2 d vD io vg],[IL1 IL2 VC1 VC2 D VD IO VG]);
B13=subs(simplify(diff(averaged_diL1_dt,d)),
    [iL1 iL2 vC1 vC2 d vD io vg],[IL1 IL2 VC1 VC2 D VD IO VG]);

B21=subs(simplify(diff(averaged_diL2_dt,io)),
    [iL1 iL2 vC1 vC2 d vD io vg],[IL1 IL2 VC1 VC2 D VD IO VG]);
B22=subs(simplify(diff(averaged_diL2_dt,vg)),
    [iL1 iL2 vC1 vC2 d vD io vg],[IL1 IL2 VC1 VC2 D VD IO VG]);
B23=subs(simplify(diff(averaged_diL2_dt,d)),
    [iL1 iL2 vC1 vC2 d vD io vg],[IL1 IL2 VC1 VC2 D VD IO VG]);

B31=subs(simplify(diff(averaged_dvC1_dt,io)),
    [iL1 iL2 vC1 vC2 d vD io vg],[IL1 IL2 VC1 VC2 D VD IO VG]);
B32=subs(simplify(diff(averaged_dvC1_dt,vg)),
    [iL1 iL2 vC1 vC2 d vD io vg],[IL1 IL2 VC1 VC2 D VD IO VG]);
B33=subs(simplify(diff(averaged_dvC1_dt,d)),
    [iL1 iL2 vC1 vC2 d vD io vg],[IL1 IL2 VC1 VC2 D VD IO VG]);

B41=subs(simplify(diff(averaged_dvC2_dt,io)),
```

```
    [iL1 iL2 vC1 vC2 d vD io vg],[IL1 IL2 VC1 VC2 D VD IO VG]);
B42=subs(simplify(diff(averaged_dvC2_dt,vg)),
    [iL1 iL2 vC1 vC2 d vD io vg],[IL1 IL2 VC1 VC2 D VD IO VG]);
B43=subs(simplify(diff(averaged_dvC2_dt,d)),
    [iL1 iL2 vC1 vC2 d vD io vg],[IL1 IL2 VC1 VC2 D VD IO VG]);

BB=eval([B11 B12 B13;
         B21 B22 B23;
         B31 B32 B33;
         B41 B42 B43]);

%Calculating the matrix C
C11=subs(simplify(diff(averaged_vo,iL1)),[iL1 iL2 vC1 vC2 d io],
    [IL1 IL2 VC1 VC2 D IO]);
C12=subs(simplify(diff(averaged_vo,iL2)),[iL1 iL2 vC1 vC2 d io],
    [IL1 IL2 VC1 VC2 D IO]);
C13=subs(simplify(diff(averaged_vo,vC1)),[iL1 iL2 vC1 vC2 d io],
    [IL1 IL2 VC1 VC2 D IO]);
C14=subs(simplify(diff(averaged_vo,vC2)),[iL1 iL2 vC1 vC2 d io],
    [IL1 IL2 VC1 VC2 D IO]);

CC=eval([C11 C12 C13 C14]);

D11=subs(simplify(diff(averaged_vo,io)),
    [iL1 iL2 vC1 vC2 d vD io vg],[IL1 IL2 VC1 VC2 D VD IO VG]);
D12=subs(simplify(diff(averaged_vo,vg)),
    [iL1 iL2 vC1 vC2 d vD io vg],[IL1 IL2 VC1 VC2 D VD IO VG]);
D13=subs(simplify(diff(averaged_vo,d)),
    [iL1 iL2 vC1 vC2 d vD io vg],[IL1 IL2 VC1 VC2 D VD IO VG]);

%Calculating the matrix D
DD=eval([D11 D12 D13]);

%Producing the State Space Model and obtaining the small
%signal transfer functions
sys=ss(AA,BB,CC,DD);
sys.inputname={'io';'vg';'d'};
sys.outputname={'vo'};
```

```matlab
vo_io=tf(sys(1,1)); %Output impedance transfer function
                    %vo(s)/io(s)
vo_vg=tf(sys(1,2)); %vo(s)/vg(s)
vo_d=tf(sys(1,3));  %Control-to-output(vo(s)/d(s))

figure(1)
bode(vo_d,{100,1e5}), hold on

figure(2)
bode(vo_vg,{100,1e5}), hold on

figure(3)
bode(vo_io,{100,1e5}), hold on
end

pause %press any key to continue
disp('Press any key to continue...')

W_vo_d=0.098207*(s+6.702e4)*(s+1579)/(s^2+1.211e4*s+8.581e7);
   %The weight obtained in previous sections for vo(s)/d(s)
W_vo_vg=0.010185*(s+8.878e5)*(s+1135)/(s^2+1.941e4*s+1.02e8);
   %The weight obtained in previous sections for vo(s)/vg(s)
W_vo_io=0.054606*(s+1.544e5)*(s+1062)/(s^2+1.808e4*s+9.748e7);
   %The weight obtained in previous sections for vo(s)/io(s)

Delta=ultidyn('Delta',[1 1]);
vo_d_unc=vo_d_nominal*(1+W_vo_d*Delta);
vo_vg_unc=vo_vg_nominal*(1+W_vo_g*Delta);
vo_io_unc=vo_io_nominal*(1+W_vo_io*Delta);

figure(1)
bode(vo_d_unc,'r--')

figure(2)
bode(vo_vg_unc,'r--')

figure(3)
bode(vo_io_unc,'r--')
```

The result shown in Figs. 2.43–2.48 are obtained after running the code.

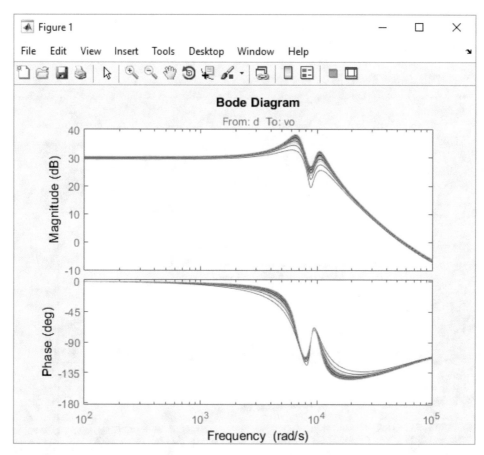

Figure 2.43: Effect of load changes on the $\frac{v_o(s)}{d(s)}$ transfer function.

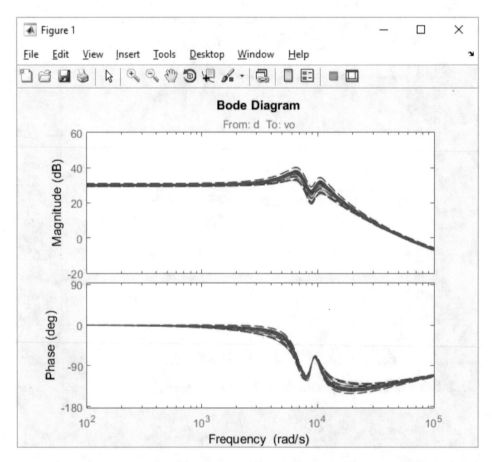

Figure 2.44: The random transfer functions produced according to the developed model, cover the transfer functions shown in Fig. 2.43.

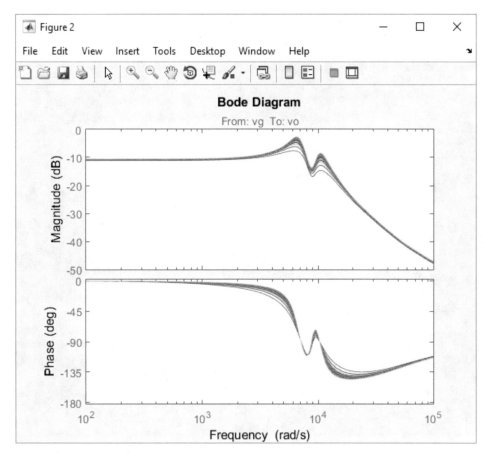

Figure 2.45: Effect of load changes on the $\frac{v_o(s)}{v_g(s)}$ transfer function.

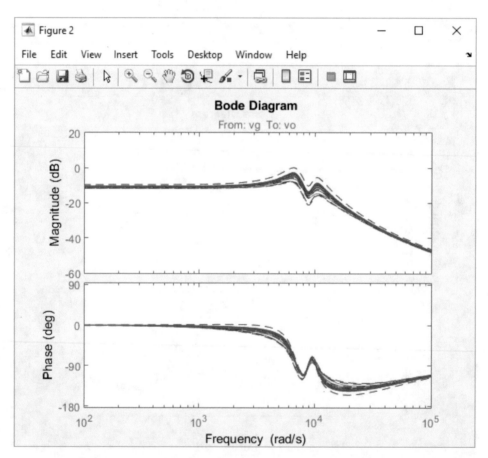

Figure 2.46: The random transfer functions produced according to the developed model cover the transfer functions shown in Fig. 2.45.

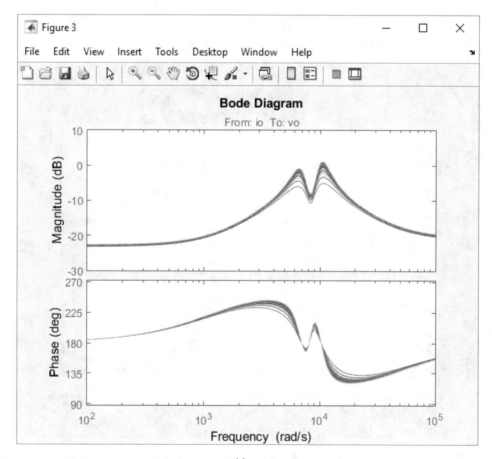

Figure 2.47: Effect of load changes on the $\frac{v_o(s)}{i_o(s)}$ transfer function.

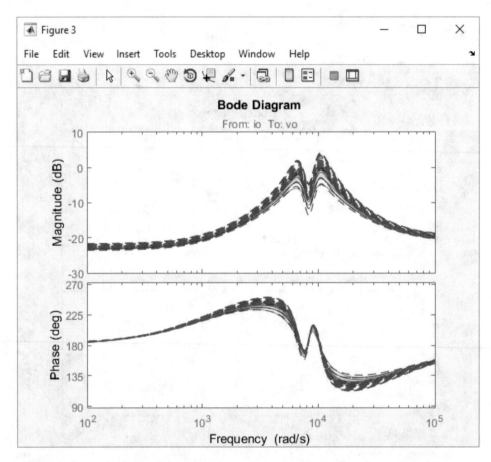

Figure 2.48: The random transfer functions produced according to the developed model cover the transfer functions shown in Fig. 2.47.

2.10 EFFECT OF COMPONENTS TOLERANCES

Previous codes study the effect of load changes only. Other components may have variations as well. This section studies the effect of components variations.

Assume the components of the Zeta converter have the variations shown in Table 2.3.

Table 2.3: The Zeta converter parameters (see Fig. 2.1)

	Nominal Value	Variations
Input DC source voltage, Vg	20 V	±20%
MOSFET Drain-Source resistance, rds	10 mΩ	±20%
Capacitor, C_1	100 μF	±20%
Capacitor C_1 Equivaluent Series Resistance (ESR), rC	0.19 Ω	[-10%,+90%]
Capacitor, C_2	220 μF	±20%
Capacitor C_2 Equivaluent Series Resistance (ESR),r C	0.095 Ω	[-10%,+90%]
Inductor, L_1	100 μH	±10%
Inductor ESR, rL_1	1 mΩ	[-10%,+90%]
Inductor, L_2	55 μH	±10%
Inductor ESR, rL_2	0.55 mΩ	[-10%,+90%]
Diode voltage drop, vD	0.7 V	±30%
Diode forward resistance, rD	10 mΩ	[-10%,+50%]
Load resistor, R	6 Ω	±80%
Switching Frequency, Fsw	100 KHz	-

The following code uses the commands provided by Robust Control Toolbox. This code studies the effect of component variations on the converter dynamics.

```
%This program calculates the small signal transfer functions
%of Zeta converter.
%This program consider the uncertainity in components.
clc
clear all

NumberOfIteration=250;
DesiredOutputVoltage=5;

n=0;
for i=1:NumberOfIteration
```

```
n=n+1;
%Definition of uncertainity in parameters
VG_unc=ureal('VG_unc',20,'Percentage',[-20 +20]);
   %Value of input DC source is in the range of 16..24
rg=0;
   %Internal resistance of input DC source
rds_unc=ureal('rds_unc',.01,'Percentage',[-20 +20]);
   %MOSFET on resistance
C1_unc=ureal('C1_unc',100e-6,'Percentage',[-20 +20]);
   %Capacitor C1 value
C2_unc=ureal('C2_unc',220e-6,'Percentage',[-20 +20]);
   %Capacitor C2 value
rC1_unc=ureal('rC1_unc',.19,'Percentage',[-10 +90]);
   %Capacitor C1 Equivalent Series Resistance(ESR)
rC2_unc=ureal('rC2_unc',.095,'Percentage',[-10 +90]);
   %Capacitor C2 Equivalent Series Resistance(ESR)
L1_unc=ureal('L1_unc',100e-6,'Percentage',[-10 +10]);
   %Inductor L1 value
L2_unc=ureal('L2_unc',55e-6,'Percentage',[-10 +10]);
   %Inductor L2 value
rL1_unc=ureal('rL1_unc',1e-3,'Percentage',[-10 +90]);
   %Inductor L1 Equivalent Series Resistance(ESR)
rL2_unc=ureal('rL2_unc',.55e-3,'Percentage',[-10 +90]);
   %Inductor L2 Equivalent Series Resistance(ESR)
rD_unc=ureal('rD_unc',.01,'Percentage',[-10 +50]);
   %Diode series resistance
VD_unc=ureal('VD_unc',.7,'Percentage',[-30 +30]);
   %Diode voltage drop
R_unc=ureal('R_unc',6,'Percentage',[-80 +80]);
   %Load resistance
IO=0;
   %Average value of output current source
fsw=100e3;
   %Switching frequency

%Sampling the uncertain set
%for instance usample(VG_unc,1) takes one sample of uncertain
%parameter VG_unc
```

```
VG=usample(VG_unc,1);
    %Sampled value of input DC source
rds=usample(rds_unc,1);
    %Sampled MOSFET on resistance
C1=usample(C1_unc,1);
    %Sampled capacitor C1 value
C2=usample(C2_unc,1);
    %Sampled capacitor C2 value
rC1=usample(rC1_unc,1);
    %Sampled capacitor C1 Equivalent Series Resistance(ESR)
rC2=usample(rC2_unc,1);
    %Sampled capacitor C2 Equivalent Series Resistance(ESR)
L1=usample(L1_unc,1);
    %Sampled inductor L1 value
L2=usample(L2_unc,1);
    %Sampled inductor L2 value
rL1=usample(rL1_unc,1);
    %Sampled inductor L1 Equivalent Series Resistance(ESR)
rL2=usample(rL2_unc,1);
    %Sampled inductor L2 Equivalent Series Resistance(ESR)
rD=usample(rD_unc,1);
    %Sampled diode series resistance
VD=usample(VD_unc,1);
    %Sampled diode voltage drop
R=usample(R_unc,1);
    %Sampled load resistance

%output voltage of an IDEAL(i.e. no losses) Zeta converter
%operating in CCM is given by:
%        D
%VO=--------VG
%       1-D
%where
%VO: average value of output voltage
%D: Duty Ratio
%VG: Input DC voltage
%So, for a IDEAL converter
%       VO
%D=----------
```

```
%      VO+VG
%Since our converter has losses we use a bigger duty ratio,
%for instance:
%            VO
%D=1.1 ----------
%            VO+VG

D=1.1*DesiredOutputVoltage/(VG+DesiredOutputVoltage);
   %Duty cylcle

syms iL1 iL2 vC1 vC2 io vg vD d
% iL1: Inductor L1 current
% iL2: Inductor L2 current
% vC1: Capacitor C1 voltage
% vC2: Capacitor C2 voltage
% io : Output current source
% vg : Input DC source
% vD : Diode voltage drop
% d  : Duty cycle

%Closed MOSFET Equations
diL1_dt_MOSFET_close=(-(rL1+rg+rds)*iL1-(rg+rds)*iL2+vg)/L1;
diL2_dt_MOSFET_close=(-(rg+rds)*iL1-(rg+rds+rC1+rL2+R*rC2/
   (R+rC2))*iL2+vC1-R/(R+rC2)*vC2+R*rC2/(R+rC2)*io+vg)/L2;
dvC1_dt_MOSFET_close=(-iL2)/C1;
dvC2_dt_MOSFET_close=(R/(R+rC2)*iL2-1/(R+rC2)*vC2-R/
   (R+rC2)*io)/C2;
vo_MOSFET_close=R*rC2/(R+rC2)*iL2+R/(R+rC2)*vC2-R*rC2/(R+rC2)*io;

%Opened MOSFET Equations
diL1_dt_MOSFET_open=(-(rL1+rC1+rD)*iL1-rD*iL2-vC1-vD)/L1;
diL2_dt_MOSFET_open=(-rD*iL1-(rD+rL2+R*rC2/(R+rC2))*iL2-R/
   (R+rC2)*vC2+R*rC2/(R+rC2)*io-vD)/L2;
dvC1_dt_MOSFET_open=(iL1)/C1;
dvC2_dt_MOSFET_open=(R/(R+rC2)*iL2-1/(R+rC2)*vC2-R/
   (R+rC2)*io)/C2;
vo_MOSFET_open=R*rC2/(R+rC2)*iL2+R/(R+rC2)*vC2-R*rC2/(R+rC2)*io;

%Averaging
```

```
averaged_diL1_dt=simplify(d*diL1_dt_MOSFET_close+(1-d)*
    diL1_dt_MOSFET_open);
averaged_diL2_dt=simplify(d*diL2_dt_MOSFET_close+(1-d)*
    diL2_dt_MOSFET_open);
averaged_dvC1_dt=simplify(d*dvC1_dt_MOSFET_close+(1-d)*
    dvC1_dt_MOSFET_open);
averaged_dvC2_dt=simplify(d*dvC2_dt_MOSFET_close+(1-d)*
    dvC2_dt_MOSFET_open);
averaged_vo=simplify(d*vo_MOSFET_close+(1-d)*vo_MOSFET_open);

%Substituting the steady values of input DC voltage source,
%Diode voltage drop, Duty cycle and output current source
%and calculating the DC operating point
right_side_of_averaged_diL1_dt=subs(averaged_diL1_dt,
    [vg vD d io],[VG VD D IO]);
right_side_of_averaged_diL2_dt=subs(averaged_diL2_dt,
    [vg vD d io],[VG VD D IO]);
right_side_of_averaged_dvC1_dt=subs(averaged_dvC1_dt,
    [vg vD d io],[VG VD D IO]);
right_side_of_averaged_dvC2_dt=subs(averaged_dvC2_dt,
    [vg vD d io],[VG VD D IO]);

DC_OPERATING_POINT=
solve(right_side_of_averaged_diL1_dt==0,
    right_side_of_averaged_diL2_dt==0,
    right_side_of_averaged_dvC1_dt==0,
    right_side_of_averaged_dvC2_dt==0,'iL1','iL2','vC1','vC2');

IL1=eval(DC_OPERATING_POINT.iL1);
IL2=eval(DC_OPERATING_POINT.iL2);
VC1=eval(DC_OPERATING_POINT.vC1);
VC2=eval(DC_OPERATING_POINT.vC2);
VO=eval(subs(averaged_vo,[iL1 iL2 vC1 vC2 io],
    [IL1 IL2 VC1 VC2 IO]));

disp('Operating point of converter')
disp('---------------------------')
disp('VO(V)=')
disp(VO)
```

```
disp('-----------------------------')

%Linearizing the averaged equations around the DC
%operating point. We want to obtain the matrix A, B, C, and D
%        .
%        x=Ax+Bu
%        y=Cx+Du
%
%where,
%        x=[iL1 iL2 vC1 vC2]'
%        u=[io vg d]'
%Since we used the variables D for steady state duty
%ratio and C to show the capacitors values we use AA, BB,
%CC and DD instead of A, B, C, and D.

%Calculating the matrix A
A11=subs(simplify(diff(averaged_diL1_dt,iL1)),
    [iL1 iL2 vC1 vC2 d io],[IL1 IL2 VC1 VC2 D IO]);
A12=subs(simplify(diff(averaged_diL1_dt,iL2)),
    [iL1 iL2 vC1 vC2 d io],[IL1 IL2 VC1 VC2 D IO]);
A13=subs(simplify(diff(averaged_diL1_dt,vC1)),
    [iL1 iL2 vC1 vC2 d io],[IL1 IL2 VC1 VC2 D IO]);
A14=subs(simplify(diff(averaged_diL1_dt,vC2)),
    [iL1 iL2 vC1 vC2 d io],[IL1 IL2 VC1 VC2 D IO]);

A21=subs(simplify(diff(averaged_diL2_dt,iL1)),
    [iL1 iL2 vC1 vC2 d io],[IL1 IL2 VC1 VC2 D IO]);
A22=subs(simplify(diff(averaged_diL2_dt,iL2)),
    [iL1 iL2 vC1 vC2 d io],[IL1 IL2 VC1 VC2 D IO]);
A23=subs(simplify(diff(averaged_diL2_dt,vC1)),
    [iL1 iL2 vC1 vC2 d io],[IL1 IL2 VC1 VC2 D IO]);
A24=subs(simplify(diff(averaged_diL2_dt,vC2)),
    [iL1 iL2 vC1 vC2 d io],[IL1 IL2 VC1 VC2 D IO]);

A31=subs(simplify(diff(averaged_dvC1_dt,iL1)),
    [iL1 iL2 vC1 vC2 d io],[IL1 IL2 VC1 VC2 D IO]);
A32=subs(simplify(diff(averaged_dvC1_dt,iL2)),
    [iL1 iL2 vC1 vC2 d io],[IL1 IL2 VC1 VC2 D IO]);
A33=subs(simplify(diff(averaged_dvC1_dt,vC1)),
```

```
      [iL1 iL2 vC1 vC2 d io],[IL1 IL2 VC1 VC2 D IO]);
A34=subs(simplify(diff(averaged_dvC1_dt,vC2)),
      [iL1 iL2 vC1 vC2 d io],[IL1 IL2 VC1 VC2 D IO]);


A41=subs(simplify(diff(averaged_dvC2_dt,iL1)),
      [iL1 iL2 vC1 vC2 d io],[IL1 IL2 VC1 VC2 D IO]);
A42=subs(simplify(diff(averaged_dvC2_dt,iL2)),
      [iL1 iL2 vC1 vC2 d io],[IL1 IL2 VC1 VC2 D IO]);
A43=subs(simplify(diff(averaged_dvC2_dt,vC1)),
      [iL1 iL2 vC1 vC2 d io],[IL1 IL2 VC1 VC2 D IO]);
A44=subs(simplify(diff(averaged_dvC2_dt,vC2)),
      [iL1 iL2 vC1 vC2 d io],[IL1 IL2 VC1 VC2 D IO]);


AA=eval([A11 A12 A13 A14;
         A21 A22 A23 A24;
         A31 A32 A33 A34;
         A41 A42 A43 A44]);


%Calculating the matrix B
B11=subs(simplify(diff(averaged_diL1_dt,io)),
      [iL1 iL2 vC1 vC2 d vD io vg],[IL1 IL2 VC1 VC2 D VD IO VG]);
B12=subs(simplify(diff(averaged_diL1_dt,vg)),
      [iL1 iL2 vC1 vC2 d vD io vg],[IL1 IL2 VC1 VC2 D VD IO VG]);
B13=subs(simplify(diff(averaged_diL1_dt,d)),
      [iL1 iL2 vC1 vC2 d vD io vg],[IL1 IL2 VC1 VC2 D VD IO VG]);


B21=subs(simplify(diff(averaged_diL2_dt,io)),
      [iL1 iL2 vC1 vC2 d vD io vg],[IL1 IL2 VC1 VC2 D VD IO VG]);
B22=subs(simplify(diff(averaged_diL2_dt,vg)),
      [iL1 iL2 vC1 vC2 d vD io vg],[IL1 IL2 VC1 VC2 D VD IO VG]);
B23=subs(simplify(diff(averaged_diL2_dt,d)),
      [iL1 iL2 vC1 vC2 d vD io vg],[IL1 IL2 VC1 VC2 D VD IO VG]);


B31=subs(simplify(diff(averaged_dvC1_dt,io)),
      [iL1 iL2 vC1 vC2 d vD io vg],[IL1 IL2 VC1 VC2 D VD IO VG]);
B32=subs(simplify(diff(averaged_dvC1_dt,vg)),
      [iL1 iL2 vC1 vC2 d vD io vg],[IL1 IL2 VC1 VC2 D VD IO VG]);
B33=subs(simplify(diff(averaged_dvC1_dt,d)),
      [iL1 iL2 vC1 vC2 d vD io vg],[IL1 IL2 VC1 VC2 D VD IO VG]);
```

```
B41=subs(simplify(diff(averaged_dvC2_dt,io)),
    [iL1 iL2 vC1 vC2 d vD io vg],[IL1 IL2 VC1 VC2 D VD IO VG]);
B42=subs(simplify(diff(averaged_dvC2_dt,vg)),
    [iL1 iL2 vC1 vC2 d vD io vg],[IL1 IL2 VC1 VC2 D VD IO VG]);
B43=subs(simplify(diff(averaged_dvC2_dt,d)),
    [iL1 iL2 vC1 vC2 d vD io vg],[IL1 IL2 VC1 VC2 D VD IO VG]);

BB=eval([B11 B12 B13;
         B21 B22 B23;
         B31 B32 B33;
         B41 B42 B43]);

%Calculating the matrix C
C11=subs(simplify(diff(averaged_vo,iL1)),[iL1 iL2 vC1 vC2 d io],
    [IL1 IL2 VC1 VC2 D IO]);
C12=subs(simplify(diff(averaged_vo,iL2)),[iL1 iL2 vC1 vC2 d io],
    [IL1 IL2 VC1 VC2 D IO]);
C13=subs(simplify(diff(averaged_vo,vC1)),[iL1 iL2 vC1 vC2 d io],
    [IL1 IL2 VC1 VC2 D IO]);
C14=subs(simplify(diff(averaged_vo,vC2)),[iL1 iL2 vC1 vC2 d io],
    [IL1 IL2 VC1 VC2 D IO]);

CC=eval([C11 C12 C13 C14]);

D11=subs(simplify(diff(averaged_vo,io)),
    [iL1 iL2 vC1 vC2 d vD io vg],[IL1 IL2 VC1 VC2 D VD IO VG]);
D12=subs(simplify(diff(averaged_vo,vg)),
    [iL1 iL2 vC1 vC2 d vD io vg],[IL1 IL2 VC1 VC2 D VD IO VG]);
D13=subs(simplify(diff(averaged_vo,d)),
    [iL1 iL2 vC1 vC2 d vD io vg],[IL1 IL2 VC1 VC2 D VD IO VG]);

%Calculating the matrix D
DD=eval([D11 D12 D13]);

%Producing the State Space Model and obtaining the small
%signal transfer functions
sys=ss(AA,BB,CC,DD);
sys.inputname={'io';'vg';'d'};
```

```
sys.outputname={'vo'};

vo_io=tf(sys(1,1)); %Output impedance transfer function
                    %vo(s)/io(s)
vo_vg=tf(sys(1,2)); %vo(s)/vg(s)
vo_d=tf(sys(1,3));  %Control-to-output(vo(s)/d(s))

%drawing the Bode diagrams
%vo_io=vo(s)/i0(s)
figure(1)
bode(vo_io),grid minor,title('vo(s)/io(s)')
hold on

%vo_vg=vo(s)/vg(s)
figure(2)
bode(vo_vg),grid minor,title('vo(s)/vg(s)')
hold on

%vo_d=vo(s)/d(s)
figure(3)
bode(vo_d),grid minor,title('vo(s)/d(s)')
hold on

%Display the progress of the loop
disp('Percentage of work done:')
disp(n/NumberOfIteration*100)
end
```

After running the code, the results shown in Figs. 2.49, 2.50, and 2.51 are obtained.

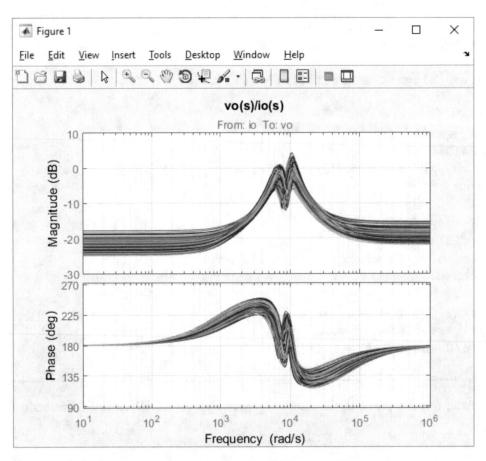

Figure 2.49: Effect of components changes on the $\frac{v_o(s)}{i_o(s)}$ transfer function.

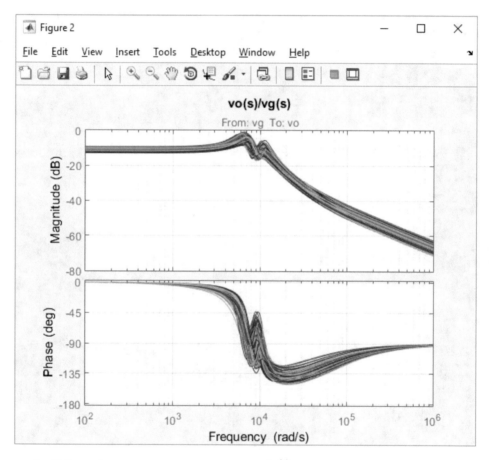

Figure 2.50: Effect of components changes on the $\frac{v_o(s)}{v_g(s)}$ transfer function.

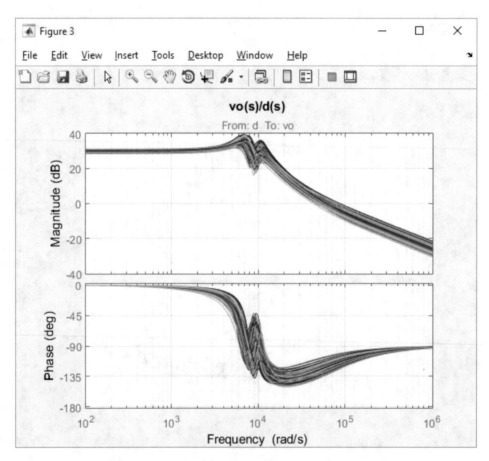

Figure 2.51: Effect of components changes on the $\frac{v_o(s)}{d(s)}$ transfer function.

2.11   OBTAINING THE UNCERTAIN MODEL OF THE CONVERTER IN PRECENCE OF COMPONENTS TOLERANCES

Although Figs. 2.49, 2.50, and 2.51 show the effect of components tolerances on the converter dynamics graphically, they give no information to form the uncertain model of the converter. The following code can be used to extract the multiplicative uncertainty model of the converter. It uses the "ucover" command to obtain the multiplicative uncertainty weight.

```
%This program calculates the multiplicative uncertainity weights
%in presence of component uncertainty.
%we use ucover command to obtain the uncertainty.
clc
clear all

NumberOfIteration=250;
DesiredOutputVoltage=5;

%Nominal transfer functions for R=Rnominal=6 ohm
%These results are obtained in the previous analysis.
s=tf('s');
DEN=(s^2+2239*s+4.76e7)*(s^2+2767*s+1.026e8);
vo_io_nominal=-.093519*(s+4.785e4)*(s+1163)*
    (s^2+1396*s+6.882e7)/DEN;
vo_vg_nominal=391.08*(s+4.785e4)*(s^2+1473*s+7.7e7)/DEN;
vo_d_nominal=43775*(s+4.785e4)*(s^2+1371*s+7.696e7)/DEN;

n=0;
for i=1:NumberOfIteration
n=n+1;
%Definition of uncertainity in parameters
VG_unc=ureal('VG_unc',20,'Percentage',[-20 +20]);
    %Value of input DC source is in the range of 16..24
rg=0;
    %Internal resistance of input DC source
rds_unc=ureal('rds_unc',.01,'Percentage',[-20 +20]);
    %MOSFET on resistance
C1_unc=ureal('C1_unc',100e-6,'Percentage',[-20 +20]);
    %Capacitor C1 value
```

```
C2_unc=ureal('C2_unc',220e-6,'Percentage',[-20 +20]);
   %Capacitor C2 value
rC1_unc=ureal('rC1_unc',.19,'Percentage',[-10 +90]);
   %Capacitor C1 Equivalent Series Resistance(ESR)
rC2_unc=ureal('rC2_unc',.095,'Percentage',[-10 +90]);
   %Capacitor C2 Equivalent Series Resistance(ESR)
 L1_unc=ureal('L1_unc',100e-6,'Percentage',[-10 +10]);
   %Inductor L1 value
L2_unc=ureal('L2_unc',55e-6,'Percentage',[-10 +10]);
   %Inductor L2 value
rL1_unc=ureal('rL1_unc',1e-3,'Percentage',[-10 +90]);
   %Inductor L1 Equivalent Series Resistance(ESR)
rL2_unc=ureal('rL2_unc',.55e-3,'Percentage',[-10 +90]);
   %Inductor L2 Equivalent Series Resistance(ESR)
rD_unc=ureal('rD_unc',.01,'Percentage',[-10 +50]);
   %Diode series resistance
VD_unc=ureal('VD_unc',.7,'Percentage',[-30 +30]);
   %Diode voltage drop
R_unc=ureal('R_unc',6,'Percentage',[-80 +80]);
   %Load resistance
IO=0;
   %Average value of output current source
fsw=100e3;
   %Switching frequency

%Sampling the uncertain set
%for instance usample(VG_unc,1) takes one sample of uncertain
%parameter VG_unc

VG=usample(VG_unc,1);
   %Sampled value of input DC source
rds=usample(rds_unc,1);
   %Sampled MOSFET on resistance
C1=usample(C1_unc,1);
   %Sampled capacitor C1 value
C2=usample(C2_unc,1);
   %Sampled capacitor C2 value
rC1=usample(rC1_unc,1);
   %Sampled capacitor C1 Equivalent Series Resistance(ESR)
```

```
rC2=usample(rC2_unc,1);
   %Sampled capacitor C2 Equivalent Series Resistance(ESR)
L1=usample(L1_unc,1);
   %Sampled inductor L1 value
L2=usample(L2_unc,1);
   %Sampled inductor L2 value
rL1=usample(rL1_unc,1);
   %Sampled inductor L1 Equivalent Series Resistance(ESR)
rL2=usample(rL2_unc,1);
   %Sampled inductor L2 Equivalent Series Resistance(ESR)
rD=usample(rD_unc,1);
   %Sampled diode series resistance
VD=usample(VD_unc,1);
   %Sampled diode voltage drop
R=usample(R_unc,1);
   %Sampled load resistance

%output voltage of an IDEAL(i.e., no losses) Zeta converter
%operating in CCM is given by:
%         D
%VO=--------VG
%        1-D
%where
%VO: average value of output voltage
%D: Duty Ratio
%VG: Input DC voltage
%So, for a IDEAL converter
%        VO
%D=----------
%     VO+VG
%Since our converter has losses we use a bigger duty ratio,
%for instance:
%           VO
%D=1.1 ----------
%         VO+VG

D=1.1*DesiredOutputVoltage/(VG+DesiredOutputVoltage);
   %Duty cylcle
```

```
syms iL1 iL2 vC1 vC2 io vg vD d
% iL1: Inductor L1 current
% iL2: Inductor L2 current
% vC1: Capacitor C1 voltage
% vC2: Capacitor C2 voltage
% io : Output current source
% vg : Input DC source
% vD : Diode voltage drop
% d  : Duty cycle

%Closed MOSFET Equations
diL1_dt_MOSFET_close=(-(rL1+rg+rds)*iL1-(rg+rds)*iL2+vg)/L1;
diL2_dt_MOSFET_close=(-(rg+rds)*iL1-(rg+rds+rC1+rL2+R*rC2/
    (R+rC2))*iL2+vC1-R/(R+rC2)*vC2+R*rC2/(R+rC2)*io+vg)/L2;
dvC1_dt_MOSFET_close=(-iL2)/C1;
dvC2_dt_MOSFET_close=(R/(R+rC2)*iL2-1/(R+rC2)*vC2-R/
    (R+rC2)*io)/C2;
vo_MOSFET_close=R*rC2/(R+rC2)*iL2+R/(R+rC2)*vC2-R*rC2/(R+rC2)*io;

%Opened MOSFET Equations
diL1_dt_MOSFET_open=(-(rL1+rC1+rD)*iL1-rD*iL2-vC1-vD)/L1;
diL2_dt_MOSFET_open=(-rD*iL1-(rD+rL2+R*rC2/(R+rC2))*iL2-R/
    (R+rC2)*vC2+R*rC2/(R+rC2)*io-vD)/L2;
dvC1_dt_MOSFET_open=(iL1)/C1;
dvC2_dt_MOSFET_open=(R/(R+rC2)*iL2-1/(R+rC2)*vC2-R/
    (R+rC2)*io)/C2;
vo_MOSFET_open=R*rC2/(R+rC2)*iL2+R/(R+rC2)*vC2-R*rC2/(R+rC2)*io;

%Averaging
averaged_diL1_dt=simplify(d*diL1_dt_MOSFET_close+(1-d)*
    diL1_dt_MOSFET_open);
averaged_diL2_dt=simplify(d*diL2_dt_MOSFET_close+(1-d)*
    diL2_dt_MOSFET_open);
averaged_dvC1_dt=simplify(d*dvC1_dt_MOSFET_close+(1-d)*
    dvC1_dt_MOSFET_open);
averaged_dvC2_dt=simplify(d*dvC2_dt_MOSFET_close+(1-d)*
    dvC2_dt_MOSFET_open);
averaged_vo=simplify(d*vo_MOSFET_close+(1-d)*vo_MOSFET_open);
```

```
%Substituting the steady values of input DC voltage source,
%Diode voltage drop, Duty cycle and output current source
%and calculating the DC operating point
right_side_of_averaged_diL1_dt=subs(averaged_diL1_dt,
    [vg vD d io],[VG VD D IO]);
right_side_of_averaged_diL2_dt=subs(averaged_diL2_dt,
    [vg vD d io],[VG VD D IO]);
right_side_of_averaged_dvC1_dt=subs(averaged_dvC1_dt,
    [vg vD d io],[VG VD D IO]);
right_side_of_averaged_dvC2_dt=subs(averaged_dvC2_dt,
    [vg vD d io],[VG VD D IO]);

DC_OPERATING_POINT=
solve(right_side_of_averaged_diL1_dt==0,
    right_side_of_averaged_diL2_dt==0,
    right_side_of_averaged_dvC1_dt==0,
    right_side_of_averaged_dvC2_dt==0,'iL1','iL2','vC1','vC2');

IL1=eval(DC_OPERATING_POINT.iL1);
IL2=eval(DC_OPERATING_POINT.iL2);
VC1=eval(DC_OPERATING_POINT.vC1);
VC2=eval(DC_OPERATING_POINT.vC2);
VO=eval(subs(averaged_vo,[iL1 iL2 vC1 vC2 io],
    [IL1 IL2 VC1 VC2 IO]));

disp('Output voltage of converter')
disp('---------------------------')
disp('VO(V)=')
disp(VO)
disp('---------------------------')

%Linearizing the averaged equations around the DC
%operating point. We want to obtain the matrix A, B, C, and D
%         .
%        x=Ax+Bu
%        y=Cx+Du
%
%where,
%        x=[iL1 iL2 vC1 vC2]'
```

```
%       u=[io vg d]'
%Since we used the variables D for steady state duty
%ratio and C to show the capacitors values we use AA,
%BB, CC and DD instead of A, B, C, and D.

%Calculating the matrix A
A11=subs(simplify(diff(averaged_diL1_dt,iL1)),
    [iL1 iL2 vC1 vC2 d io],[IL1 IL2 VC1 VC2 D IO]);
A12=subs(simplify(diff(averaged_diL1_dt,iL2)),
    [iL1 iL2 vC1 vC2 d io],[IL1 IL2 VC1 VC2 D IO]);
A13=subs(simplify(diff(averaged_diL1_dt,vC1)),
    [iL1 iL2 vC1 vC2 d io],[IL1 IL2 VC1 VC2 D IO]);
A14=subs(simplify(diff(averaged_diL1_dt,vC2)),
    [iL1 iL2 vC1 vC2 d io],[IL1 IL2 VC1 VC2 D IO]);

A21=subs(simplify(diff(averaged_diL2_dt,iL1)),
    [iL1 iL2 vC1 vC2 d io],[IL1 IL2 VC1 VC2 D IO]);
A22=subs(simplify(diff(averaged_diL2_dt,iL2)),
    [iL1 iL2 vC1 vC2 d io],[IL1 IL2 VC1 VC2 D IO]);
A23=subs(simplify(diff(averaged_diL2_dt,vC1)),
    [iL1 iL2 vC1 vC2 d io],[IL1 IL2 VC1 VC2 D IO]);
A24=subs(simplify(diff(averaged_diL2_dt,vC2)),
    [iL1 iL2 vC1 vC2 d io],[IL1 IL2 VC1 VC2 D IO]);

A31=subs(simplify(diff(averaged_dvC1_dt,iL1)),
    [iL1 iL2 vC1 vC2 d io],[IL1 IL2 VC1 VC2 D IO]);
A32=subs(simplify(diff(averaged_dvC1_dt,iL2)),
    [iL1 iL2 vC1 vC2 d io],[IL1 IL2 VC1 VC2 D IO]);
A33=subs(simplify(diff(averaged_dvC1_dt,vC1)),
    [iL1 iL2 vC1 vC2 d io],[IL1 IL2 VC1 VC2 D IO]);
A34=subs(simplify(diff(averaged_dvC1_dt,vC2)),
    [iL1 iL2 vC1 vC2 d io],[IL1 IL2 VC1 VC2 D IO]);

A41=subs(simplify(diff(averaged_dvC2_dt,iL1)),
    [iL1 iL2 vC1 vC2 d io],[IL1 IL2 VC1 VC2 D IO]);
A42=subs(simplify(diff(averaged_dvC2_dt,iL2)),
    [iL1 iL2 vC1 vC2 d io],[IL1 IL2 VC1 VC2 D IO]);
A43=subs(simplify(diff(averaged_dvC2_dt,vC1)),
    [iL1 iL2 vC1 vC2 d io],[IL1 IL2 VC1 VC2 D IO]);
```

```
A44=subs(simplify(diff(averaged_dvC2_dt,vC2)),
   [iL1 iL2 vC1 vC2 d io],[IL1 IL2 VC1 VC2 D IO]);


AA=eval([A11 A12 A13 A14;
         A21 A22 A23 A24;
         A31 A32 A33 A34;
         A41 A42 A43 A44]);

%Calculating the matrix B
B11=subs(simplify(diff(averaged_diL1_dt,io)),
   [iL1 iL2 vC1 vC2 d vD io vg],[IL1 IL2 VC1 VC2 D VD IO VG]);
B12=subs(simplify(diff(averaged_diL1_dt,vg)),
   [iL1 iL2 vC1 vC2 d vD io vg],[IL1 IL2 VC1 VC2 D VD IO VG]);
B13=subs(simplify(diff(averaged_diL1_dt,d)),
   [iL1 iL2 vC1 vC2 d vD io vg],[IL1 IL2 VC1 VC2 D VD IO VG]);

B21=subs(simplify(diff(averaged_diL2_dt,io)),
   [iL1 iL2 vC1 vC2 d vD io vg],[IL1 IL2 VC1 VC2 D VD IO VG]);
B22=subs(simplify(diff(averaged_diL2_dt,vg)),
   [iL1 iL2 vC1 vC2 d vD io vg],[IL1 IL2 VC1 VC2 D VD IO VG]);
B23=subs(simplify(diff(averaged_diL2_dt,d)),
   [iL1 iL2 vC1 vC2 d vD io vg],[IL1 IL2 VC1 VC2 D VD IO VG]);

B31=subs(simplify(diff(averaged_dvC1_dt,io)),
   [iL1 iL2 vC1 vC2 d vD io vg],[IL1 IL2 VC1 VC2 D VD IO VG]);
B32=subs(simplify(diff(averaged_dvC1_dt,vg)),
   [iL1 iL2 vC1 vC2 d vD io vg],[IL1 IL2 VC1 VC2 D VD IO VG]);
B33=subs(simplify(diff(averaged_dvC1_dt,d)),
   [iL1 iL2 vC1 vC2 d vD io vg],[IL1 IL2 VC1 VC2 D VD IO VG]);

B41=subs(simplify(diff(averaged_dvC2_dt,io)),
   [iL1 iL2 vC1 vC2 d vD io vg],[IL1 IL2 VC1 VC2 D VD IO VG]);
B42=subs(simplify(diff(averaged_dvC2_dt,vg)),
   [iL1 iL2 vC1 vC2 d vD io vg],[IL1 IL2 VC1 VC2 D VD IO VG]);
B43=subs(simplify(diff(averaged_dvC2_dt,d)),
   [iL1 iL2 vC1 vC2 d vD io vg],[IL1 IL2 VC1 VC2 D VD IO VG]);

BB=eval([B11 B12 B13;
         B21 B22 B23;
```

```
        B31 B32 B33;
        B41 B42 B43]);

%Calculating the matrix C
C11=subs(simplify(diff(averaged_vo,iL1)),[iL1 iL2 vC1 vC2 d io],
    [IL1 IL2 VC1 VC2 D IO]);
C12=subs(simplify(diff(averaged_vo,iL2)),[iL1 iL2 vC1 vC2 d io],
    [IL1 IL2 VC1 VC2 D IO]);
C13=subs(simplify(diff(averaged_vo,vC1)),[iL1 iL2 vC1 vC2 d io],
    [IL1 IL2 VC1 VC2 D IO]);
C14=subs(simplify(diff(averaged_vo,vC2)),[iL1 iL2 vC1 vC2 d io],
    [IL1 IL2 VC1 VC2 D IO]);

CC=eval([C11 C12 C13 C14]);

D11=subs(simplify(diff(averaged_vo,io)),
    [iL1 iL2 vC1 vC2 d vD io vg],[IL1 IL2 VC1 VC2 D VD IO VG]);
D12=subs(simplify(diff(averaged_vo,vg)),
    [iL1 iL2 vC1 vC2 d vD io vg],[IL1 IL2 VC1 VC2 D VD IO VG]);
D13=subs(simplify(diff(averaged_vo,d)),
    [iL1 iL2 vC1 vC2 d vD io vg],[IL1 IL2 VC1 VC2 D VD IO VG]);

%Calculating the matrix D
DD=eval([D11 D12 D13]);

%Producing the State Space Model and obtaining the small
%signal transfer functions
sys=ss(AA,BB,CC,DD);
sys.inputname={'io';'vg';'d'};
sys.outputname={'vo'};

vo_io=tf(sys(1,1)); %Output impedance transfer function
                    %vo(s)/io(s)
vo_vg=tf(sys(1,2)); %vo(s)/vg(s)
vo_d=tf(sys(1,3));  %Control-to-output(vo(s)/d(s))

if n==1
    %there is no variable names "array" in the first
    %running of the loop.
```

```
    %variable "array" is initialize in the first
    %running of loop.
    array1=vo_d;
    array2=vo_vg;
    array3=vo_io;
else
    array1=stack(1,array1,vo_d);
    array2=stack(1,array2,vo_vg);
    array3=stack(1,array3,vo_io);
end
disp('Percentage of work done:')
disp(n/NumberOfIteration*100) %shows the progress of the loop
disp('')
end

%Calculating the multiplicative uncertainity bounds
omega=logspace(-1,5,200);
array1_frd=frd(array1,omega);
array2_frd=frd(array2,omega);
array3_frd=frd(array3,omega);

relerr1 = (vo_d_nominal-array1_frd)/vo_d_nominal;
relerr2 = (vo_vg_nominal-array2_frd)/vo_vg_nominal;
relerr3 = (vo_io_nominal-array3_frd)/vo_io_nominal;

[P1,Info1]=ucover(array1_frd,vo_d_nominal,2);
    %fitting a 2nd order weight to vo(s)/d(s).
[P2,Info2]=ucover(array2_frd,vo_vg_nominal,2);
    %fitting a 2nd order weight to vo(s)/vg(s).
[P3,Info3]=ucover(array3_frd,vo_io_nominal,2);
    %fitting a 2nd order weight to vo(s)/io(s).

figure(1)
bodemag(relerr1,'b--',Info1.W1,'r',{0.1,100000});
figure(2)
bodemag(relerr2,'b--',Info2.W1,'r',{0.1,100000});
figure(3)
bodemag(relerr3,'b--',Info3.W1,'r',{0.1,100000});
```

After running the code, the results shown in Figs. 2.52, 2.53, and 2.54 are obtained.

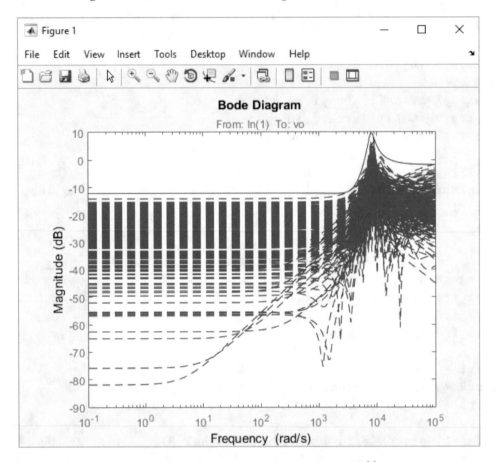

Figure 2.52: Obtaining the multiplicative uncertainty weights for $\frac{v_o(s)}{d(s)}$.

One can obtain the calculated weights using the commands shown in Figs. 2.55, 2.56, and 2.57.

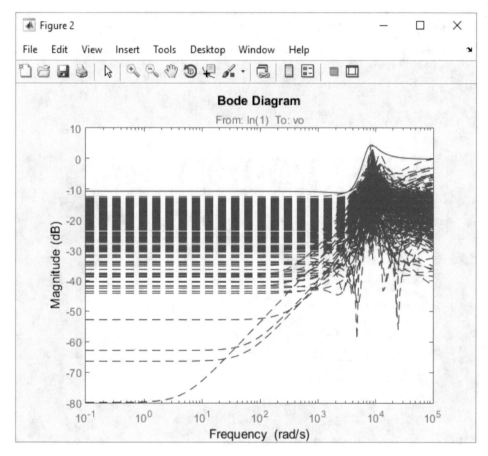

Figure 2.53: Obtaining the multiplicative uncertainty weights for $\frac{v_o(s)}{v_g(s)}$.

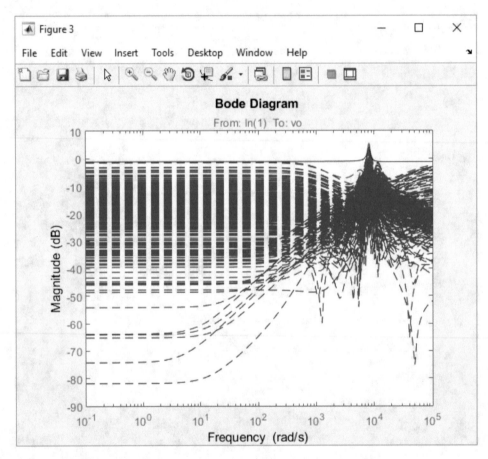

Figure 2.54: Obtaining the multiplicative uncertainty weights for $\frac{v_o(s)}{i_o(s)}$.

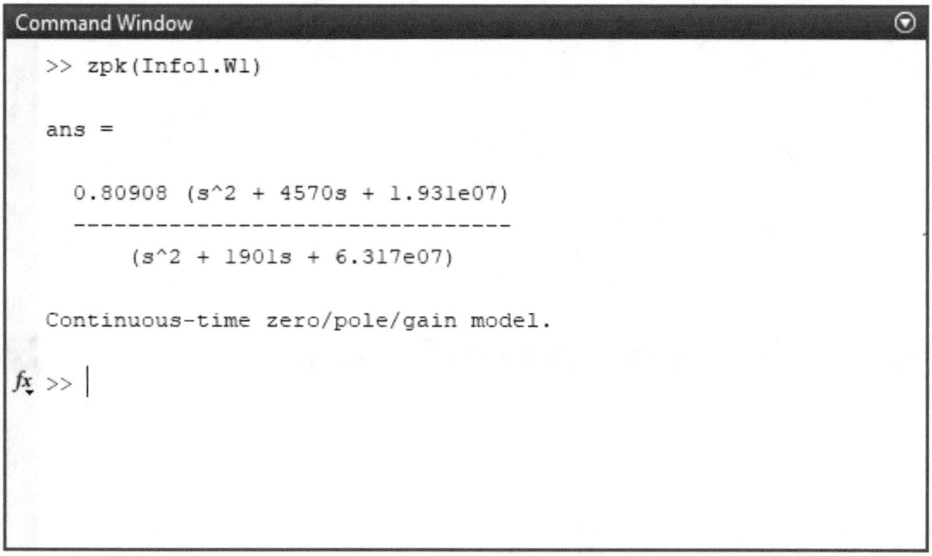

```
Command Window                                                          ⊙

    >> zpk(Info1.W1)

    ans =

      0.80908 (s^2 + 4570s + 1.931e07)
      --------------------------------
          (s^2 + 1901s + 6.317e07)

    Continuous-time zero/pole/gain model.

fx >> |
```

Figure 2.55: Equation of obtained multiplicative uncertainty weight for $\frac{v_o(s)}{d(s)}$.

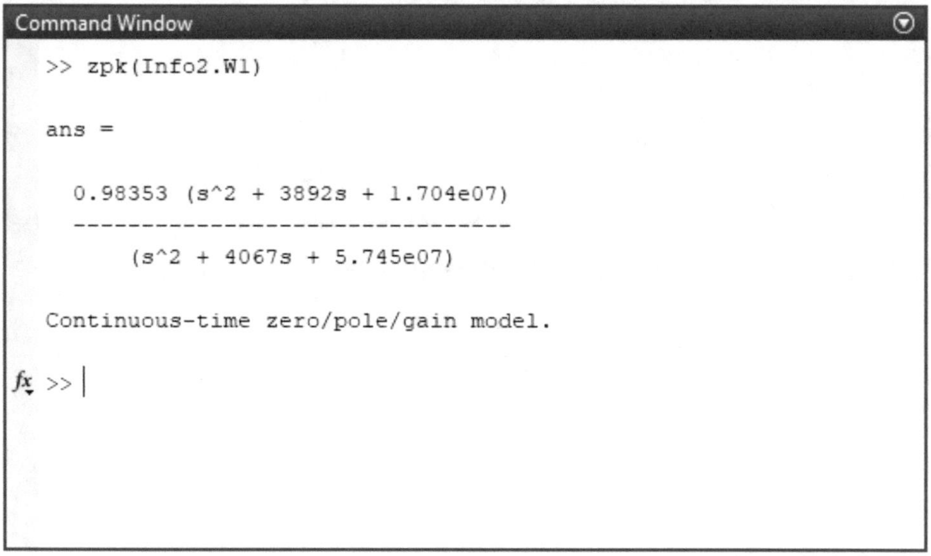

```
Command Window                                                          ⊙

    >> zpk(Info2.W1)

    ans =

      0.98353 (s^2 + 3892s + 1.704e07)
      --------------------------------
          (s^2 + 4067s + 5.745e07)

    Continuous-time zero/pole/gain model.

fx >> |
```

Figure 2.56: Equation of obtained multiplicative uncertainty weight for $\frac{v_o(s)}{v_g(s)}$.

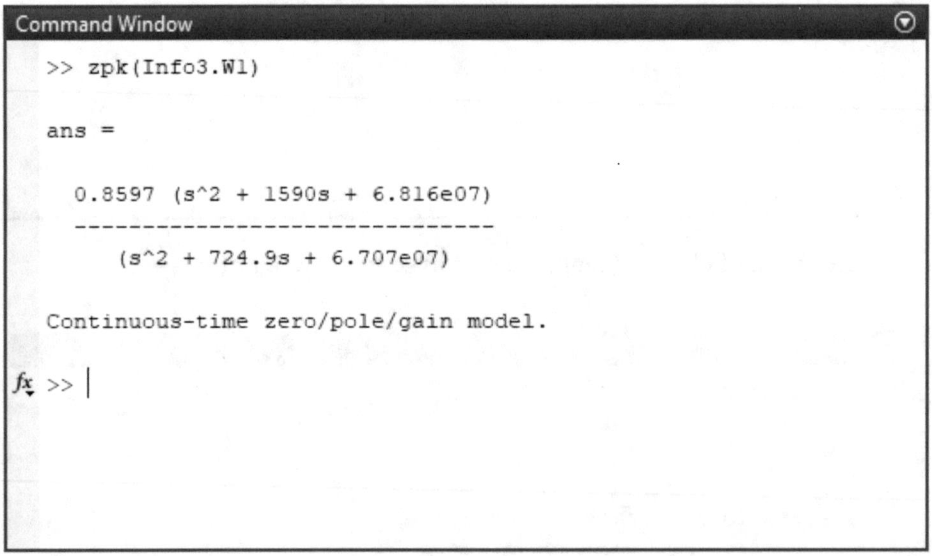

```
Command Window                                                    ⊙

 >> zpk(Info3.W1)

 ans =

   0.8597 (s^2 + 1590s + 6.816e07)
   -------------------------------
      (s^2 + 724.9s + 6.707e07)

 Continuous-time zero/pole/gain model.

fx >> |
```

Figure 2.57: Equation of obtained multiplicative uncertainty weight for $\frac{v_o(s)}{i_o(s)}$.

2.12   TESTING THE OBTAINED UNCERTAINTY WEIGHTS

The reliability of obtained model can be tested similar to Section 2.9. The following code can be used to learn whether or not the obtained model is reliable.

```
%This program shows the goodness of obtained weights
clc
clear all

NumberOfIteration=250;
DesiredOutputVoltage=5;

%Nominal transfer functions for R=Rnominal=6 ohm
%These results are obtained in the previous analysis.
s=tf('s');
DEN=(s^2+2239*s+4.76e7)*(s^2+2767*s+1.026e8);
vo_io_nominal=-.093519*(s+4.785e4)*(s+1163)*
    (s^2+1396*s+6.882e7)/DEN;
vo_vg_nominal=391.08*(s+4.785e4)*(s^2+1473*s+7.7e7)/DEN;
vo_d_nominal=43775*(s+4.785e4)*(s^2+1371*s+7.696e7)/DEN;

n=0;
for i=1:NumberOfIteration
n=n+1;
%Definition of uncertainity in parameters
VG_unc=ureal('VG_unc',20,'Percentage',[-20 +20]);
    %Value of input DC source is in the range of 16..24
rg=0;
    %Internal resistance of input DC source
rds_unc=ureal('rds_unc',.01,'Percentage',[-20 +20]);
    %MOSFET on resistance
C1_unc=ureal('C1_unc',100e-6,'Percentage',[-20 +20]);
    %Capacitor C1 value
C2_unc=ureal('C2_unc',220e-6,'Percentage',[-20 +20]);
    %Capacitor C2 value
rC1_unc=ureal('rC1_unc',.19,'Percentage',[-10 +90]);
    %Capacitor C1 Equivalent Series Resistance(ESR)
rC2_unc=ureal('rC2_unc',.095,'Percentage',[-10 +90]);
    %Capacitor C2 Equivalent Series Resistance(ESR)
L1_unc=ureal('L1_unc',100e-6,'Percentage',[-10 +10]);
```

```
    %Inductor L1 value
L2_unc=ureal('L2_unc',55e-6,'Percentage',[-10 +10]);
    %Inductor L2 value
rL1_unc=ureal('rL1_unc',1e-3,'Percentage',[-10 +90]);
    %Inductor L1 Equivalent Series Resistance(ESR)
rL2_unc=ureal('rL2_unc',.55e-3,'Percentage',[-10 +90]);
    %Inductor L2 Equivalent Series Resistance(ESR)
rD_unc=ureal('rD_unc',.01,'Percentage',[-10 +50]);
    %Diode series resistance
VD_unc=ureal('VD_unc',.7,'Percentage',[-30 +30]);
    %Diode voltage drop
R_unc=ureal('R_unc',6,'Percentage',[-80 +80]);
    %Load resistance
IO=0;
    %Average value of output current source
fsw=100e3;
    %Switching frequency

%Sampling the uncertain set
%for instance usample(VG_unc,1) takes one sample of uncertain
%parameter VG_unc

VG=usample(VG_unc,1);
    %Sampled value of input DC source
rds=usample(rds_unc,1);
    %Sampled MOSFET on resistance
C1=usample(C1_unc,1);
    %Sampled capacitor C1 value
C2=usample(C2_unc,1);
    %Sampled capacitor C2 value
rC1=usample(rC1_unc,1);
    %Sampled capacitor C1 Equivalent Series Resistance(ESR)
rC2=usample(rC2_unc,1);
    %Sampled capacitor C2 Equivalent Series Resistance(ESR)
L1=usample(L1_unc,1);
    %Sampled inductor L1 value
L2=usample(L2_unc,1);
    %Sampled inductor L2 value
rL1=usample(rL1_unc,1);
```

```
   %Sampled inductor L1 Equivalent Series Resistance(ESR)
rL2=usample(rL2_unc,1);
   %Sampled inductor L2 Equivalent Series Resistance(ESR)
rD=usample(rD_unc,1);
   %Sampled diode series resistance
VD=usample(VD_unc,1);
   %Sampled diode voltage drop
R=usample(R_unc,1);
   %Sampled load resistance

%output voltage of an IDEAL(i.e., no losses) Zeta
%converter operating in CCM is given by:
%         D
%VO=--------VG
%        1-D
%where
%VO: average value of output voltage
%D: Duty Ratio
%VG: Input DC voltage
%So, for a IDEAL converter
%        VO
%D=----------
%     VO+VG
%Since our converter has losses we use a bigger duty ratio,
%for instance:
%             VO
%D=1.1 ----------
%          VO+VG

D=1.1*DesiredOutputVoltage/(VG+DesiredOutputVoltage);
   %Duty cylcle

syms iL1 iL2 vC1 vC2 io vg vD d
% iL1: Inductor L1 current
% iL2: Inductor L2 current
% vC1: Capacitor C1 voltage
% vC2: Capacitor C2 voltage
% io : Output current source
% vg : Input DC source
```

```
% vD : Diode voltage drop
% d  : Duty cycle

%Closed MOSFET Equations
diL1_dt_MOSFET_close=(-(rL1+rg+rds)*iL1-(rg+rds)*iL2+vg)/L1;
diL2_dt_MOSFET_close=(-(rg+rds)*iL1-(rg+rds+rC1+rL2+R*rC2/
   (R+rC2))*iL2+vC1-R/(R+rC2)*vC2+R*rC2/(R+rC2)*io+vg)/L2;
dvC1_dt_MOSFET_close=(-iL2)/C1;
dvC2_dt_MOSFET_close=(R/(R+rC2)*iL2-1/(R+rC2)*vC2-R/
   (R+rC2)*io)/C2;
vo_MOSFET_close=R*rC2/(R+rC2)*iL2+R/(R+rC2)*vC2-R*rC2/(R+rC2)*io;

%Opened MOSFET Equations
diL1_dt_MOSFET_open=(-(rL1+rC1+rD)*iL1-rD*iL2-vC1-vD)/L1;
diL2_dt_MOSFET_open=(-rD*iL1-(rD+rL2+R*rC2/(R+rC2))*iL2-R/
   (R+rC2)*vC2+R*rC2/(R+rC2)*io-vD)/L2;
dvC1_dt_MOSFET_open=(iL1)/C1;
dvC2_dt_MOSFET_open=(R/(R+rC2)*iL2-1/(R+rC2)*vC2-R/
   (R+rC2)*io)/C2;
vo_MOSFET_open=R*rC2/(R+rC2)*iL2+R/(R+rC2)*vC2-R*rC2/(R+rC2)*io;

%Averaging
averaged_diL1_dt=simplify(d*diL1_dt_MOSFET_close+(1-d)*
   diL1_dt_MOSFET_open);
averaged_diL2_dt=simplify(d*diL2_dt_MOSFET_close+(1-d)*
   diL2_dt_MOSFET_open);
averaged_dvC1_dt=simplify(d*dvC1_dt_MOSFET_close+(1-d)*
   dvC1_dt_MOSFET_open);
averaged_dvC2_dt=simplify(d*dvC2_dt_MOSFET_close+(1-d)*
   dvC2_dt_MOSFET_open);
averaged_vo=simplify(d*vo_MOSFET_close+(1-d)*vo_MOSFET_open);

%Substituting the steady values of input DC voltage source,
%Diode voltage drop, Duty cycle and output current source
%and calculating the DC operating point
right_side_of_averaged_diL1_dt=subs(averaged_diL1_dt,
   [vg vD d io],[VG VD D IO]);
right_side_of_averaged_diL2_dt=subs(averaged_diL2_dt,
   [vg vD d io],[VG VD D IO]);
```

```
right_side_of_averaged_dvC1_dt=subs(averaged_dvC1_dt,
    [vg vD d io],[VG VD D IO]);
right_side_of_averaged_dvC2_dt=subs(averaged_dvC2_dt,
    [vg vD d io],[VG VD D IO]);

DC_OPERATING_POINT=
solve(right_side_of_averaged_diL1_dt==0,
    right_side_of_averaged_diL2_dt==0,
    right_side_of_averaged_dvC1_dt==0,
    right_side_of_averaged_dvC2_dt==0,'iL1','iL2','vC1','vC2');

IL1=eval(DC_OPERATING_POINT.iL1);
IL2=eval(DC_OPERATING_POINT.iL2);
VC1=eval(DC_OPERATING_POINT.vC1);
VC2=eval(DC_OPERATING_POINT.vC2);
VO=eval(subs(averaged_vo,[iL1 iL2 vC1 vC2 io],
    [IL1 IL2 VC1 VC2 IO]));

disp('Output voltage of converter')
disp('---------------------------')
disp('VO(V)=')
disp(VO)
disp('---------------------------')

%Linearizing the averaged equations around the DC
%operating point. We want to obtain the matrix A, B, C, and D
%          .
%       x=Ax+Bu
%       y=Cx+Du
%
%where,
%       x=[iL1 iL2 vC1 vC2]'
%       u=[io vg d]'
%Since we used the variables D for steady state duty
%ratio and C to show the capacitors values we use AA,
%BB, CC and DD instead of A, B, C, and D.

%Calculating the matrix A
A11=subs(simplify(diff(averaged_diL1_dt,iL1)),
```

```
      [iL1 iL2 vC1 vC2 d io],[IL1 IL2 VC1 VC2 D IO]);
 A12=subs(simplify(diff(averaged_diL1_dt,iL2)),
      [iL1 iL2 vC1 vC2 d io],[IL1 IL2 VC1 VC2 D IO]);
 A13=subs(simplify(diff(averaged_diL1_dt,vC1)),
      [iL1 iL2 vC1 vC2 d io],[IL1 IL2 VC1 VC2 D IO]);
 A14=subs(simplify(diff(averaged_diL1_dt,vC2)),
      [iL1 iL2 vC1 vC2 d io],[IL1 IL2 VC1 VC2 D IO]);

 A21=subs(simplify(diff(averaged_diL2_dt,iL1)),
      [iL1 iL2 vC1 vC2 d io],[IL1 IL2 VC1 VC2 D IO]);
 A22=subs(simplify(diff(averaged_diL2_dt,iL2)),
      [iL1 iL2 vC1 vC2 d io],[IL1 IL2 VC1 VC2 D IO]);
 A23=subs(simplify(diff(averaged_diL2_dt,vC1)),
      [iL1 iL2 vC1 vC2 d io],[IL1 IL2 VC1 VC2 D IO]);
 A24=subs(simplify(diff(averaged_diL2_dt,vC2)),
      [iL1 iL2 vC1 vC2 d io],[IL1 IL2 VC1 VC2 D IO]);

 A31=subs(simplify(diff(averaged_dvC1_dt,iL1)),
      [iL1 iL2 vC1 vC2 d io],[IL1 IL2 VC1 VC2 D IO]);
 A32=subs(simplify(diff(averaged_dvC1_dt,iL2)),
      [iL1 iL2 vC1 vC2 d io],[IL1 IL2 VC1 VC2 D IO]);
 A33=subs(simplify(diff(averaged_dvC1_dt,vC1)),
      [iL1 iL2 vC1 vC2 d io],[IL1 IL2 VC1 VC2 D IO]);
 A34=subs(simplify(diff(averaged_dvC1_dt,vC2)),
      [iL1 iL2 vC1 vC2 d io],[IL1 IL2 VC1 VC2 D IO]);

 A41=subs(simplify(diff(averaged_dvC2_dt,iL1)),
      [iL1 iL2 vC1 vC2 d io],[IL1 IL2 VC1 VC2 D IO]);
 A42=subs(simplify(diff(averaged_dvC2_dt,iL2)),
      [iL1 iL2 vC1 vC2 d io],[IL1 IL2 VC1 VC2 D IO]);
 A43=subs(simplify(diff(averaged_dvC2_dt,vC1)),
      [iL1 iL2 vC1 vC2 d io],[IL1 IL2 VC1 VC2 D IO]);
 A44=subs(simplify(diff(averaged_dvC2_dt,vC2)),
      [iL1 iL2 vC1 vC2 d io],[IL1 IL2 VC1 VC2 D IO]);

 AA=eval([A11 A12 A13 A14;
          A21 A22 A23 A24;
          A31 A32 A33 A34;
          A41 A42 A43 A44]);
```

```
%Calculating the matrix B
B11=subs(simplify(diff(averaged_diL1_dt,io)),
   [iL1 iL2 vC1 vC2 d vD io vg],[IL1 IL2 VC1 VC2 D VD IO VG]);
B12=subs(simplify(diff(averaged_diL1_dt,vg)),
   [iL1 iL2 vC1 vC2 d vD io vg],[IL1 IL2 VC1 VC2 D VD IO VG]);
B13=subs(simplify(diff(averaged_diL1_dt,d)),
   [iL1 iL2 vC1 vC2 d vD io vg],[IL1 IL2 VC1 VC2 D VD IO VG]);

B21=subs(simplify(diff(averaged_diL2_dt,io)),
   [iL1 iL2 vC1 vC2 d vD io vg],[IL1 IL2 VC1 VC2 D VD IO VG]);
B22=subs(simplify(diff(averaged_diL2_dt,vg)),
   [iL1 iL2 vC1 vC2 d vD io vg],[IL1 IL2 VC1 VC2 D VD IO VG]);
B23=subs(simplify(diff(averaged_diL2_dt,d)),
   [iL1 iL2 vC1 vC2 d vD io vg],[IL1 IL2 VC1 VC2 D VD IO VG]);

B31=subs(simplify(diff(averaged_dvC1_dt,io)),
   [iL1 iL2 vC1 vC2 d vD io vg],[IL1 IL2 VC1 VC2 D VD IO VG]);
B32=subs(simplify(diff(averaged_dvC1_dt,vg)),
   [iL1 iL2 vC1 vC2 d vD io vg],[IL1 IL2 VC1 VC2 D VD IO VG]);
B33=subs(simplify(diff(averaged_dvC1_dt,d)),
   [iL1 iL2 vC1 vC2 d vD io vg],[IL1 IL2 VC1 VC2 D VD IO VG]);

B41=subs(simplify(diff(averaged_dvC2_dt,io)),
   [iL1 iL2 vC1 vC2 d vD io vg],[IL1 IL2 VC1 VC2 D VD IO VG]);
B42=subs(simplify(diff(averaged_dvC2_dt,vg)),
   [iL1 iL2 vC1 vC2 d vD io vg],[IL1 IL2 VC1 VC2 D VD IO VG]);
B43=subs(simplify(diff(averaged_dvC2_dt,d)),
   [iL1 iL2 vC1 vC2 d vD io vg],[IL1 IL2 VC1 VC2 D VD IO VG]);

BB=eval([B11 B12 B13;
         B21 B22 B23;
         B31 B32 B33;
         B41 B42 B43]);

%Calculating the matrix C
C11=subs(simplify(diff(averaged_vo,iL1)),[iL1 iL2 vC1 vC2 d io],
   [IL1 IL2 VC1 VC2 D IO]);
C12=subs(simplify(diff(averaged_vo,iL2)),[iL1 iL2 vC1 vC2 d io],
```

```
       [IL1 IL2 VC1 VC2 D IO]);
C13=subs(simplify(diff(averaged_vo,vC1)),[iL1 iL2 vC1 vC2 d io],
       [IL1 IL2 VC1 VC2 D IO]);
C14=subs(simplify(diff(averaged_vo,vC2)),[iL1 iL2 vC1 vC2 d io],
       [IL1 IL2 VC1 VC2 D IO]);


CC=eval([C11 C12 C13 C14]);


D11=subs(simplify(diff(averaged_vo,io)),
       [iL1 iL2 vC1 vC2 d vD io vg],[IL1 IL2 VC1 VC2 D VD IO VG]);
D12=subs(simplify(diff(averaged_vo,vg)),
       [iL1 iL2 vC1 vC2 d vD io vg],[IL1 IL2 VC1 VC2 D VD IO VG]);
D13=subs(simplify(diff(averaged_vo,d)),
       [iL1 iL2 vC1 vC2 d vD io vg],[IL1 IL2 VC1 VC2 D VD IO VG]);


%Calculating the matrix D
DD=eval([D11 D12 D13]);


%Producing the State Space Model and obtaining the
%small signal transfer functions
sys=ss(AA,BB,CC,DD);
sys.inputname={'io';'vg';'d'};
sys.outputname={'vo'};


vo_io=tf(sys(1,1)); %Output impedance transfer function
                    %vo(s)/io(s)
vo_vg=tf(sys(1,2)); %vo(s)/vg(s)
vo_d=tf(sys(1,3));  %Control-to-output(vo(s)/d(s))


%drawing the Bode diagrams


%vo_d=vo(s)/d(s)
figure(1)
bode(vo_d),grid minor,title('vo(s)/d(s)')
hold on


%vo_vg=vo(s)/vg(s)
figure(2)
bode(vo_vg),grid minor,title('vo(s)/vg(s)')
```

```
hold on

%vo_io=vo(s)/i0(s)
figure(3)
bode(vo_io),grid minor,title('vo(s)/io(s)')
hold on

disp('Percentage of work done:')
disp(n/NumberOfIteration*100) %shows the progress of the loop
disp('')
end

pause
disp('Press any key to continue...')

W_vo_d=.80908*(s^2+4570*s+1.931e7)/(s^2+1901*s+6.317e7);
   %The weight obtained in previous sections for vo(s)/d(s)
W_vo_vg=.98353*(s^2+3892*s+1.704e7)/(s^2+4067*s+5.745e7);
   %The weight obtained in previous sections for vo(s)/vg(s)
W_vo_io=.8597*(s^2+1590*s+6.816e7)/(s^2+724.9*s+6.707e7);
   %The weight obtained in previous sections for vo(s)/io(s)

Delta=ultidyn('Delta',[1 1]);
vo_d_unc=vo_d_nominal*(1+W_vo_d*Delta);
vo_vg_unc=vo_vg_nominal*(1+W_vo_vg*Delta);
vo_io_unc=vo_io_nominal*(1+W_vo_io*Delta);

figure(1)
bode(vo_d_unc,'r--')

figure(2)
bode(vo_vg_unc,'r--')

figure(3)
bode(vo_io_unc,'r--')
```

After running the code, the results shown in Figs. 2.58, 2.59, and 2.60 will appear. The program asks the user to press any key to start the second part of the code.

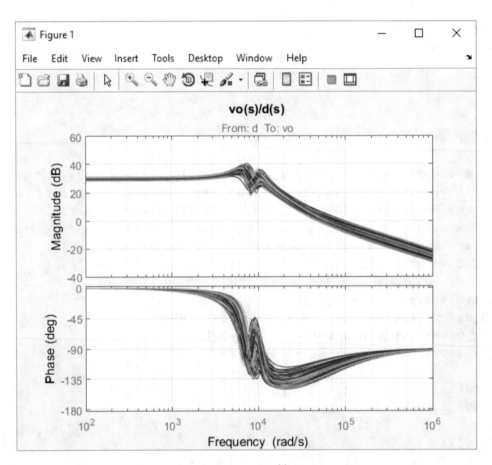

Figure 2.58: Effect of components changes on the $\frac{v_o(s)}{d(s)}$ transfer function.

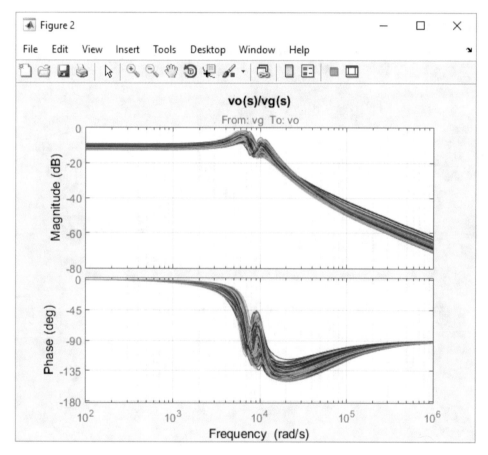

Figure 2.59: Effect of components changes on the $\frac{v_o(s)}{v_g(s)}$ transfer function.

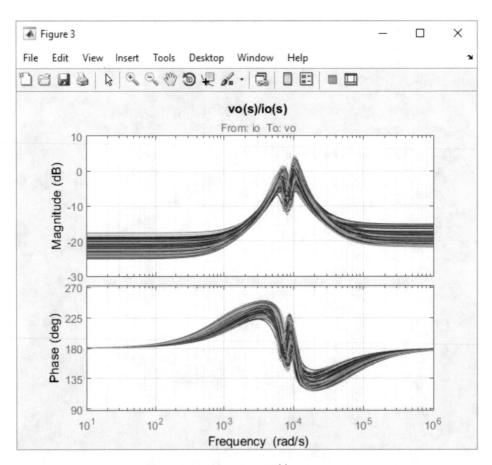

Figure 2.60: Effect of components changes on the $\frac{v_o(s)}{i_o(s)}$ transfer function.

After pressing a key, the second part of the code will run. The results shown in Figs. 2.61, 2.62, and 2.63 show the reliability of the uncertain model since it covers the results produced by the first part of the code.

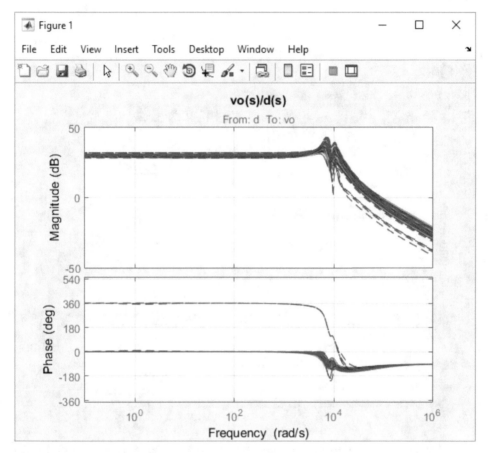

Figure 2.61: The random transfer functions produced according to the developed model cover the transfer functions shown in Fig. 2.58.

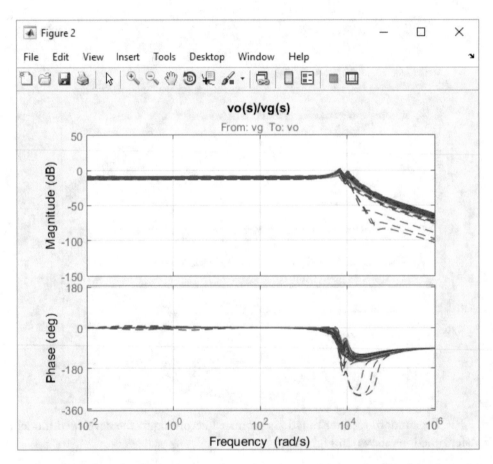

Figure 2.62: The random transfer functions produced according to the developed model cover the transfer functions shown in Fig. 2.59.

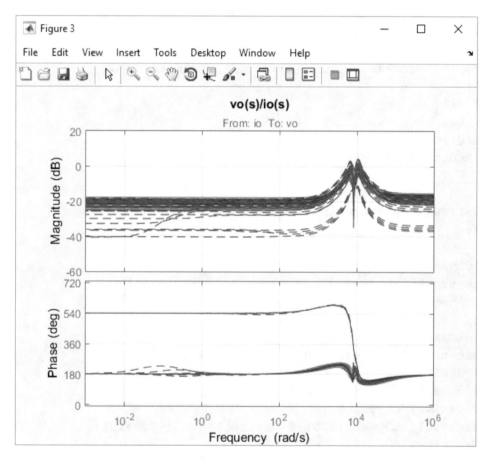

Figure 2.63: The random transfer functions produced according to the developed model cover the transfer functions shown in Fig. 2.60.

2.13 CALCULATING THE MAXIMUM/MINIMUM OF THE TRANSFER FUNCTION COEFFICIENTS

The following code extracts the interval plant model (i.e., maximum and minimum values of the transfer function coefficients in presence of variations) of the converter. The variation is the one shown in Table 2.3.

```
%This program calculates the changes in transfer
%function coefficients it draws the pzmap of transfer
%functions for varios values of components as well
clc
clear all

NumberOfIteration=25;
DesiredOutputVoltage=5;

n=0;
for i=1:NumberOfIteration
n=n+1;

%Definition of uncertainity in parameters
VG_unc=ureal('VG_unc',20,'Percentage',[-20 +20]);
    %Value of input DC source is in the range of 16..24
rg=0;
    %Internal resistance of input DC source
rds_unc=ureal('rds_unc',.01,'Percentage',[-20 +20]);
    %MOSFET on resistance
C1_unc=ureal('C1_unc',100e-6,'Percentage',[-20 +20]);
    %Capacitor C1 value
C2_unc=ureal('C2_unc',220e-6,'Percentage',[-20 +20]);
    %Capacitor C2 value
rC1_unc=ureal('rC1_unc',.19,'Percentage',[-10 +90]);
    %Capacitor C1 Equivalent Series Resistance(ESR)
rC2_unc=ureal('rC2_unc',.095,'Percentage',[-10 +90]);
    %Capacitor C2 Equivalent Series Resistance(ESR)
L1_unc=ureal('L1_unc',100e-6,'Percentage',[-10 +10]);
    %Inductor L1 value
L2_unc=ureal('L2_unc',55e-6,'Percentage',[-10 +10]);
    %Inductor L2 value
rL1_unc=ureal('rL1_unc',1e-3,'Percentage',[-10 +90]);
```

```
   %Inductor L1 Equivalent Series Resistance(ESR)
rL2_unc=ureal('rL2_unc',.55e-3,'Percentage',[-10 +90]);
   %Inductor L2 Equivalent Series Resistance(ESR)
rD_unc=ureal('rD_unc',.01,'Percentage',[-10 +50]);
   %Diode series resistance
VD_unc=ureal('VD_unc',.7,'Percentage',[-30 +30]);
   %Diode voltage drop
R_unc=ureal('R_unc',6,'Percentage';[-80 +80]);
   %Load resistance
IO=0;
   %Average value of output current source
fsw=100e3;
   %Switching frequency

%Sampling the uncertain set
%for instance usample(VG_unc,1) takes one sample of uncertain
%parameter VG_unc

VG=usample(VG_unc,1);
   %Sampled value of input DC source
rds=usample(rds_unc,1);
   %Sampled MOSFET on resistance
C1=usample(C1_unc,1);
   %Sampled capacitor C1 value
C2=usample(C2_unc,1);
   %Sampled capacitor C2 value
rC1=usample(rC1_unc,1);
   %Sampled capacitor C1 Equivalent Series Resistance(ESR)
rC2=usample(rC2_unc,1);
   %Sampled capacitor C2 Equivalent Series Resistance(ESR)
L1=usample(L1_unc,1);
   %Sampled inductor L1 value
L2=usample(L2_unc,1);
   %Sampled inductor L2 value
rL1=usample(rL1_unc,1);
   %Sampled inductor L1 Equivalent Series Resistance(ESR)
rL2=usample(rL2_unc,1);
   %Sampled inductor L2 Equivalent Series Resistance(ESR)
rD=usample(rD_unc,1);
```

```
    %Sampled diode series resistance
VD=usample(VD_unc,1);
    %Sampled diode voltage drop
R=usample(R_unc,1);
    %Sampled load resistance

%output voltage of an IDEAL(i.e. no losses) Zeta converter
%operating in CCM is given by:
%         D
%VO=--------VG
%        1-D
%where
%VO: average value of output voltage
%D: Duty Ratio
%VG: Input DC voltage
%So, for a IDEAL converter
%        VO
%D=----------
%    VO+VG
%Since our converter has losses we use a bigger duty ratio,
%for instance:
%              VO
%D=1.1 ----------
%            VO+VG

D=1.1*DesiredOutputVoltage/(VG+DesiredOutputVoltage);
    %Duty cylcle

syms iL1 iL2 vC1 vC2 io vg vD d
% iL1: Inductor L1 current
% iL2: Inductor L2 current
% vC1: Capacitor C1 voltage
% vC2: Capacitor C2 voltage
% io : Output current source
% vg : Input DC source
% vD : Diode voltage drop
% d  : Duty cycle

%Closed MOSFET Equations
```

```
diL1_dt_MOSFET_close=(-(rL1+rg+rds)*iL1-(rg+rds)*iL2+vg)/L1;
diL2_dt_MOSFET_close=(-(rg+rds)*iL1-(rg+rds+rC1+rL2+R*rC2/
   (R+rC2))*iL2+vC1-R/(R+rC2)*vC2+R*rC2/(R+rC2)*io+vg)/L2;
dvC1_dt_MOSFET_close=(-iL2)/C1;
dvC2_dt_MOSFET_close=(R/(R+rC2)*iL2-1/(R+rC2)*vC2-R/
   (R+rC2)*io)/C2;
vo_MOSFET_close=R*rC2/(R+rC2)*iL2+R/(R+rC2)*vC2-R*rC2/(R+rC2)*io;

%Opened MOSFET Equations
diL1_dt_MOSFET_open=(-(rL1+rC1+rD)*iL1-rD*iL2-vC1-vD)/L1;
diL2_dt_MOSFET_open=(-rD*iL1-(rD+rL2+R*rC2/(R+rC2))*iL2-R/
   (R+rC2)*vC2+R*rC2/(R+rC2)*io-vD)/L2;
dvC1_dt_MOSFET_open=(iL1)/C1;
dvC2_dt_MOSFET_open=(R/(R+rC2)*iL2-1/(R+rC2)*vC2-R/
   (R+rC2)*io)/C2;
vo_MOSFET_open=R*rC2/(R+rC2)*iL2+R/(R+rC2)*vC2-R*rC2/(R+rC2)*io;

%Averaging
averaged_diL1_dt=simplify(d*diL1_dt_MOSFET_close+(1-d)*
   diL1_dt_MOSFET_open);
averaged_diL2_dt=simplify(d*diL2_dt_MOSFET_close+(1-d)*
   diL2_dt_MOSFET_open);
averaged_dvC1_dt=simplify(d*dvC1_dt_MOSFET_close+(1-d)*
   dvC1_dt_MOSFET_open);
averaged_dvC2_dt=simplify(d*dvC2_dt_MOSFET_close+(1-d)*
   dvC2_dt_MOSFET_open);
averaged_vo=simplify(d*vo_MOSFET_close+(1-d)*vo_MOSFET_open);

%Substituting the steady values of input DC voltage source,
%Diode voltage drop, Duty cycle and output current source
%and calculating the DC operating point
right_side_of_averaged_diL1_dt=subs(averaged_diL1_dt,
   [vg vD d io],[VG VD D IO]);
right_side_of_averaged_diL2_dt=subs(averaged_diL2_dt,
   [vg vD d io],[VG VD D IO]);
right_side_of_averaged_dvC1_dt=subs(averaged_dvC1_dt,
   [vg vD d io],[VG VD D IO]);
right_side_of_averaged_dvC2_dt=subs(averaged_dvC2_dt,
   [vg vD d io],[VG VD D IO]);
```

```
DC_OPERATING_POINT=
solve(right_side_of_averaged_diL1_dt==0,
   right_side_of_averaged_diL2_dt==0,
   right_side_of_averaged_dvC1_dt==0,
   right_side_of_averaged_dvC2_dt==0,'iL1','iL2','vC1','vC2');

IL1=eval(DC_OPERATING_POINT.iL1);
IL2=eval(DC_OPERATING_POINT.iL2);
VC1=eval(DC_OPERATING_POINT.vC1);
VC2=eval(DC_OPERATING_POINT.vC2);
VO=eval(subs(averaged_vo,[iL1 iL2 vC1 vC2 io],
   [IL1 IL2 VC1 VC2 IO]));

disp('Operating point of converter')
disp('----------------------------')
disp('VO(V)=')
disp(VO)
disp('----------------------------')

%Linearizing the averaged equations around the DC
%operating point. We want to obtain the matrix A, B, C, and D
%        .
%        x=Ax+Bu
%        y=Cx+Du
%
%where,
%        x=[iL1 iL2 vC1 vC2]'
%        u=[io vg d]'
%Since we used the variables D for steady state duty
%ratio and C to show the capacitors values we use AA,
%BB, CC and DD instead of A, B, C, and D.

%Calculating the matrix A
A11=subs(simplify(diff(averaged_diL1_dt,iL1)),
   [iL1 iL2 vC1 vC2 d io],[IL1 IL2 VC1 VC2 D IO]);
A12=subs(simplify(diff(averaged_diL1_dt,iL2)),
   [iL1 iL2 vC1 vC2 d io],[IL1 IL2 VC1 VC2 D IO]);
A13=subs(simplify(diff(averaged_diL1_dt,vC1)),
```

```
      [iL1 iL2 vC1 vC2 d io],[IL1 IL2 VC1 VC2 D IO]);
A14=subs(simplify(diff(averaged_diL1_dt,vC2)),
      [iL1 iL2 vC1 vC2 d io],[IL1 IL2 VC1 VC2 D IO]);

A21=subs(simplify(diff(averaged_diL2_dt,iL1)),
      [iL1 iL2 vC1 vC2 d io],[IL1 IL2 VC1 VC2 D IO]);
A22=subs(simplify(diff(averaged_diL2_dt,iL2)),
      [iL1 iL2 vC1 vC2 d io],[IL1 IL2 VC1 VC2 D IO]);
A23=subs(simplify(diff(averaged_diL2_dt,vC1)),
      [iL1 iL2 vC1 vC2 d io],[IL1 IL2 VC1 VC2 D IO]);
A24=subs(simplify(diff(averaged_diL2_dt,vC2)),
      [iL1 iL2 vC1 vC2 d io],[IL1 IL2 VC1 VC2 D IO]);

A31=subs(simplify(diff(averaged_dvC1_dt,iL1)),
      [iL1 iL2 vC1 vC2 d io],[IL1 IL2 VC1 VC2 D IO]);
A32=subs(simplify(diff(averaged_dvC1_dt,iL2)),
      [iL1 iL2 vC1 vC2 d io],[IL1 IL2 VC1 VC2 D IO]);
A33=subs(simplify(diff(averaged_dvC1_dt,vC1)),
      [iL1 iL2 vC1 vC2 d io],[IL1 IL2 VC1 VC2 D IO]);
A34=subs(simplify(diff(averaged_dvC1_dt,vC2)),
      [iL1 iL2 vC1 vC2 d io],[IL1 IL2 VC1 VC2 D IO]);

A41=subs(simplify(diff(averaged_dvC2_dt,iL1)),
      [iL1 iL2 vC1 vC2 d io],[IL1 IL2 VC1 VC2 D IO]);
A42=subs(simplify(diff(averaged_dvC2_dt,iL2)),
      [iL1 iL2 vC1 vC2 d io],[IL1 IL2 VC1 VC2 D IO]);
A43=subs(simplify(diff(averaged_dvC2_dt,vC1)),
      [iL1 iL2 vC1 vC2 d io],[IL1 IL2 VC1 VC2 D IO]);
A44=subs(simplify(diff(averaged_dvC2_dt,vC2)),
      [iL1 iL2 vC1 vC2 d io],[IL1 IL2 VC1 VC2 D IO]);

AA=eval([A11 A12 A13 A14;
         A21 A22 A23 A24;
         A31 A32 A33 A34;
         A41 A42 A43 A44]);

%Calculating the matrix B
B11=subs(simplify(diff(averaged_diL1_dt,io)),
    [iL1 iL2 vC1 vC2 d vD io vg],[IL1 IL2 VC1 VC2 D VD IO VG]);
```

```
B12=subs(simplify(diff(averaged_diL1_dt,vg)),
    [iL1 iL2 vC1 vC2 d vD io vg],[IL1 IL2 VC1 VC2 D VD IO VG]);
B13=subs(simplify(diff(averaged_diL1_dt,d)),
    [iL1 iL2 vC1 vC2 d vD io vg],[IL1 IL2 VC1 VC2 D VD IO VG]);

B21=subs(simplify(diff(averaged_diL2_dt,io)),
    [iL1 iL2 vC1 vC2 d vD io vg],[IL1 IL2 VC1 VC2 D VD IO VG]);
B22=subs(simplify(diff(averaged_diL2_dt,vg)),
    [iL1 iL2 vC1 vC2 d vD io vg],[IL1 IL2 VC1 VC2 D VD IO VG]);
B23=subs(simplify(diff(averaged_diL2_dt,d)),
    [iL1 iL2 vC1 vC2 d vD io vg],[IL1 IL2 VC1 VC2 D VD IO VG]);

B31=subs(simplify(diff(averaged_dvC1_dt,io)),
    [iL1 iL2 vC1 vC2 d vD io vg],[IL1 IL2 VC1 VC2 D VD IO VG]);
B32=subs(simplify(diff(averaged_dvC1_dt,vg)),
    [iL1 iL2 vC1 vC2 d vD io vg],[IL1 IL2 VC1 VC2 D VD IO VG]);
B33=subs(simplify(diff(averaged_dvC1_dt,d)),
    [iL1 iL2 vC1 vC2 d vD io vg],[IL1 IL2 VC1 VC2 D VD IO VG]);

B41=subs(simplify(diff(averaged_dvC2_dt,io)),
    [iL1 iL2 vC1 vC2 d vD io vg],[IL1 IL2 VC1 VC2 D VD IO VG]);
B42=subs(simplify(diff(averaged_dvC2_dt,vg)),
    [iL1 iL2 vC1 vC2 d vD io vg],[IL1 IL2 VC1 VC2 D VD IO VG]);
B43=subs(simplify(diff(averaged_dvC2_dt,d)),
    [iL1 iL2 vC1 vC2 d vD io vg],[IL1 IL2 VC1 VC2 D VD IO VG]);

BB=eval([B11 B12 B13;
         B21 B22 B23;
         B31 B32 B33;
         B41 B42 B43]);

%Calculating the matrix C
C11=subs(simplify(diff(averaged_vo,iL1)),[iL1 iL2 vC1 vC2 d io],
    [IL1 IL2 VC1 VC2 D IO]);
C12=subs(simplify(diff(averaged_vo,iL2)),[iL1 iL2 vC1 vC2 d io],
    [IL1 IL2 VC1 VC2 D IO]);
C13=subs(simplify(diff(averaged_vo,vC1)),[iL1 iL2 vC1 vC2 d io],
    [IL1 IL2 VC1 VC2 D IO]);
C14=subs(simplify(diff(averaged_vo,vC2)),[iL1 iL2 vC1 vC2 d io],
```

```
   [IL1 IL2 VC1 VC2 D IO]);

CC=eval([C11 C12 C13 C14]);

D11=subs(simplify(diff(averaged_vo,io)),
   [iL1 iL2 vC1 vC2 d vD io vg],[IL1 IL2 VC1 VC2 D VD IO VG]);
D12=subs(simplify(diff(averaged_vo,vg)),
   [iL1 iL2 vC1 vC2 d vD io vg],[IL1 IL2 VC1 VC2 D VD IO VG]);
D13=subs(simplify(diff(averaged_vo,d)),
   [iL1 iL2 vC1 vC2 d vD io vg],[IL1 IL2 VC1 VC2 D VD IO VG]);

%Calculating the matrix D
DD=eval([D11 D12 D13]);

%Producing the State Space Model and obtaining the small
%signal transfer functions
sys=ss(AA,BB,CC,DD);
sys.inputname={'io';'vg';'d'};
sys.outputname={'vo'};

vo_io=tf(sys(1,1)); %Output impedance transfer function
                    %vo(s)/io(s)
vo_vg=tf(sys(1,2)); %vo(s)/vg(s)
vo_d=tf(sys(1,3));  %Control-to-output(vo(s)/d(s))

%Drawing the pole-zero maps
figure(1)
pzmap(vo_io);
hold on

figure(2)
pzmap(vo_vg);
hold on

figure(3)
pzmap(vo_d);
hold on

%Extracts the transfer function coefficients
```

```matlab
if n==1
        [num_vo_io,den_vo_io]=tfdata(vo_io,'v');
        [num_vo_vg,den_vo_vg]=tfdata(vo_vg,'v');
        [num_vo_d,den_vo_d]=tfdata(vo_d,'v');
else
        [num1,den1]=tfdata(vo_io,'v');
        num_vo_io=[num_vo_io;num1];
        den_vo_io=[den_vo_io;den1];

        [num2,den2]=tfdata(vo_vg,'v');
        num_vo_vg=[num_vo_vg;num2];
        den_vo_vg=[den_vo_vg;den2];

        [num3,den3]=tfdata(vo_d,'v');
        num_vo_d=[num_vo_d;num3];
        den_vo_d=[den_vo_d;den3];
end
%Display the progress of the loop
disp('Percentage of work done:')
disp(n/NumberOfIteration*100)
end
%Title of windows
%we put them outside of the loop since there is no need to
%run them in each iteration. This speeds up the simulations.
figure(1)
grid on
title('Pole-zero map of vo(s)/io(s)')

figure(2)
grid on
title('Pole-zero map of vo(s)/vg(s)')

figure(3)
grid on
title('Pole-zero map of vo(s)/d(s)')

%Calculates the bounds for coefficients
disp('Results')
disp('-------')
```

```
%Calculate the bounds of vo(s)/io(s)
disp('                 4    3    2           ')
disp(' vo(s)     b4*s +b3*s +b2*s +b1*s+b0')
disp('-------= ----------------------------')
disp(' io(s)       4    3    2           ')
disp('          a4*s +a3*s +a2*s +a1*s+a0')

disp('Min and Max. of b4 (vo(s)/io(s))=')
disp([min(num_vo_io(:,1)) max(num_vo_io(:,1))])

disp('Min and Max. of b3 (vo(s)/io(s))=')
disp([min(num_vo_io(:,2)) max(num_vo_io(:,2))])

disp('Min and Max. of b2 (vo(s)/io(s))=')
disp([min(num_vo_io(:,3)) max(num_vo_io(:,3))])

disp('Min and Max. of b1 (vo(s)/io(s))=')
disp([min(num_vo_io(:,4)) max(num_vo_io(:,4))])

disp('Min and Max. of b0 (vo(s)/io(s))=')
disp([min(num_vo_io(:,5)) max(num_vo_io(:,5))])

disp('Min and Max. of a4 (vo(s)/io(s))=')
disp([min(den_vo_io(:,1)) max(den_vo_io(:,1))])

disp('Min and Max. of a3 (vo(s)/io(s))=')
disp([min(den_vo_io(:,2)) max(den_vo_io(:,2))])

disp('Min and Max. of a2 (vo(s)/io(s))=')
disp([min(den_vo_io(:,3)) max(den_vo_io(:,3))])

disp('Min and Max. of a1 (vo(s)/io(s))=')
disp([min(den_vo_io(:,4)) max(den_vo_io(:,4))])

disp('Min and Max. of a0 (vo(s)/io(s))=')
disp([min(den_vo_io(:,5)) max(den_vo_io(:,5))])
disp('---------------------------------------------')
disp('Press any key ...')
pause
```

```
%Calculate the bounds of vo(s)/vg(s)
disp('               4      3       2              ')
disp(' vo(s)     b4*s +b3*s +b2*s +b1*s+b0')
disp('-------= ----------------------------')
disp(' vg(s)        4      3       2              ')
disp('            a4*s +a3*s +a2*s +a1*s+a0')

disp('Min and Max. of b4 (vo(s)/vg(s))=')
disp([min(num_vo_vg(:,1)) max(num_vo_vg(:,1))])

disp('Min and Max. of b3 (vo(s)/vg(s))=')
disp([min(num_vo_vg(:,2)) max(num_vo_vg(:,2))])

disp('Min and Max. of b2 (vo(s)/vg(s))=')
disp([min(num_vo_vg(:,3)) max(num_vo_vg(:,3))])

disp('Min and Max. of b1 (vo(s)/vg(s))=')
disp([min(num_vo_vg(:,4)) max(num_vo_vg(:,4))])

disp('Min and Max. of b0 (vo(s)/vg(s))=')
disp([min(num_vo_vg(:,5)) max(num_vo_vg(:,5))])

disp('Min and Max. of a4 (vo(s)/vg(s))=')
disp([min(den_vo_vg(:,1)) max(den_vo_vg(:,1))])

disp('Min and Max. of a3 (vo(s)/vg(s))=')
disp([min(den_vo_vg(:,2)) max(den_vo_vg(:,2))])

disp('Min and Max. of a2 (vo(s)/vg(s))=')
disp([min(den_vo_vg(:,3)) max(den_vo_vg(:,3))])

disp('Min and Max. of a1 (vo(s)/vg(s))=')
disp([min(den_vo_vg(:,4)) max(den_vo_vg(:,4))])

disp('Min and Max. of a0 (vo(s)/vg(s))=')
disp([min(den_vo_vg(:,5)) max(den_vo_vg(:,5))])
disp('-------------------------------------------')
disp('Press any key ...')
```

```
pause

%Calculate the bounds of vo(s)/d(s)
disp('                 4    3    2          ')
disp(' vo(s)     b4*s +b3*s +b2*s +b1*s+b0')
disp('-------= ---------------------------')
disp(' d(s)           4    3    2          ')
disp('          a4*s +a3*s +a2*s +a1*s+a0')

disp('Min and Max. of b4 (vo(s)/d(s))=')
disp([min(num_vo_d(:,1)) max(num_vo_d(:,1))])

disp('Min and Max. of b3 (vo(s)/d(s))=')
disp([min(num_vo_d(:,2)) max(num_vo_d(:,2))])

disp('Min and Max. of b2 (vo(s)/d(s))=')
disp([min(num_vo_d(:,3)) max(num_vo_d(:,3))])

disp('Min and Max. of b1 (vo(s)/d(s))=')
disp([min(num_vo_d(:,4)) max(num_vo_d(:,4))])

disp('Min and Max. of b0 (vo(s)/d(s))=')
disp([min(num_vo_d(:,5)) max(num_vo_d(:,5))])

disp('Min and Max. of a4 (vo(s)/d(s))=')
disp([min(den_vo_d(:,1)) max(den_vo_d(:,1))])

disp('Min and Max. of a3 (vo(s)/d(s))=')
disp([min(den_vo_d(:,2)) max(den_vo_d(:,2))])

disp('Min and Max. of a2 (vo(s)/d(s))=')
disp([min(den_vo_d(:,3)) max(den_vo_d(:,3))])

disp('Min and Max. of a1 (vo(s)/d(s))=')
disp([min(den_vo_d(:,4)) max(den_vo_d(:,4))])

disp('Min and Max. of a0 (vo(s)/d(s))=')
disp([min(den_vo_d(:,5)) max(den_vo_d(:,5))])
disp('-------------------------------------------')
```

After running the code, the results shown in Figs. 2.64–2.69 are obtained.

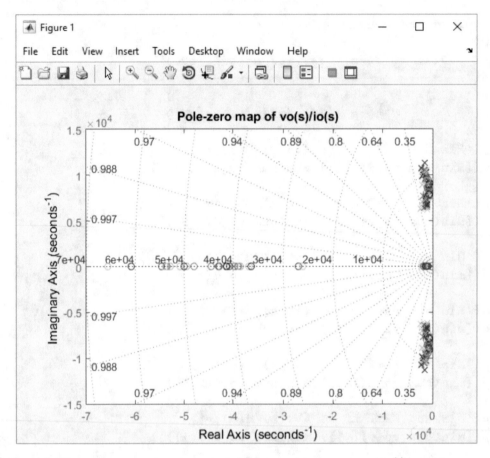

Figure 2.64: Effect of component changes on the zero and poles of the $\frac{v_o(s)}{i_o(s)}$ transfer function.

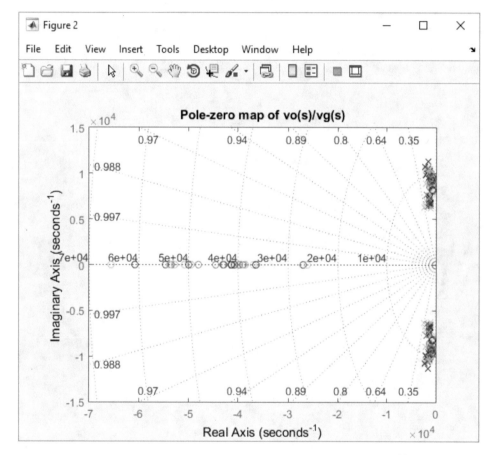

Figure 2.65: Effect of component changes on the zero and poles of the $\frac{v_o(s)}{v_g(s)}$ transfer function.

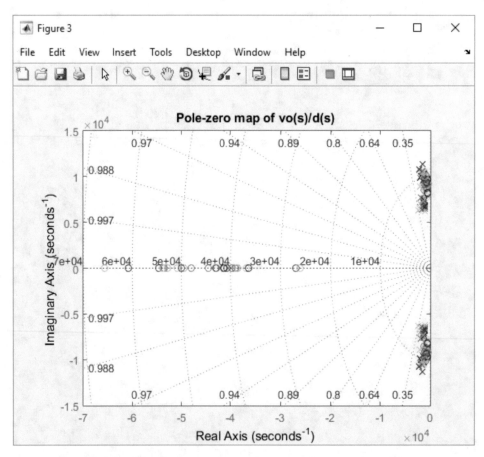

Figure 2.66: Effect of component changes on the zero and poles of the $\frac{v_o(s)}{d(s)}$ transfer function.

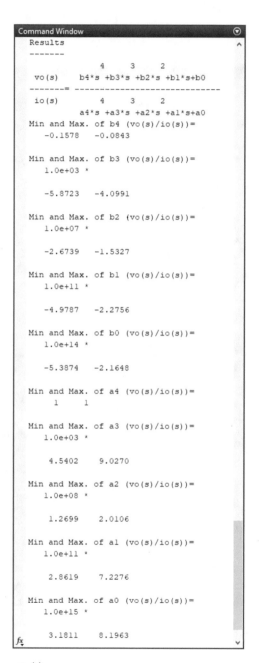

Figure 2.67: Change of the $\frac{v_o(s)}{i_o(s)}$ transfer function coefficients.

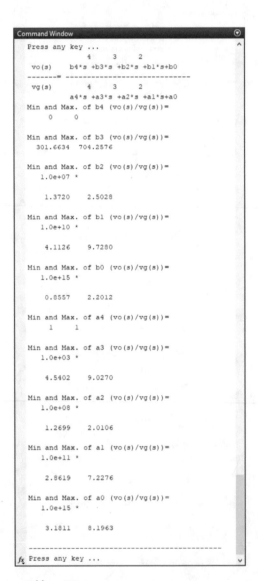

Figure 2.68: Change of the $\frac{v_o(s)}{v_g(s)}$ transfer function coefficients.

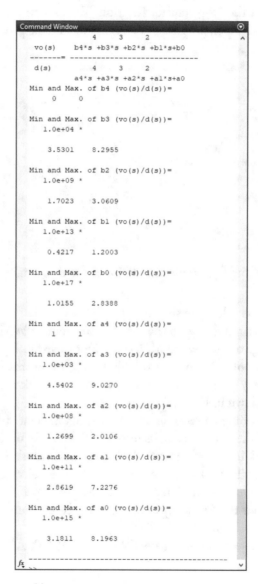

Figure 2.69: Change of the $\frac{v_o(s)}{d(s)}$ transfer function coefficients.

2.14   ANALYZING THE SYSTEM WITHOUT UNCERTAINTY

PLECS®can be used to extract the dynamic of DC-DC converters. The schematic shown in Fig. 2.70 is used to extract the $\frac{v_o(s)}{d(s)}$ transfer function. The user adds the "Small Signal Pertur-bation" and "Small Signal Responce" blocks (Fig. 2.71) to suitable places in order to extract the desired transfer function.

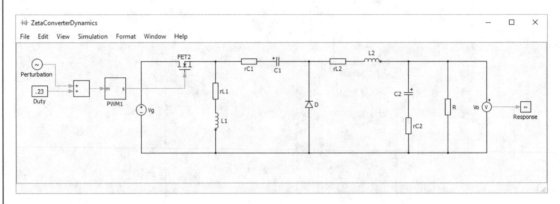

Figure 2.70: Extraction of the $\frac{v_o(s)}{d(s)}$ transfer function Bode diagram.

The used components values are shown in Fig. 2.72.

The PWM signal is produced with the aid of a "Sawtooth PWM" block (Fig. 2.73). Fig-ure 2.74 shows the settings of the "Sawtooth PWM" block used in Fig. 2.70.

Click the "Analysis tools…" in order to obtain the converter small signal transfer function. Enter the desired frequency range in the opened window and click the "Start analysis" button. The simulation result is shown in Fig. 2.77.

You can export the obtained result as a .csv (Comma Seperated Values) file. The .csv file is the medium to transfer the obtained results into MATLAB®. In order to export the obtained graphic as an .csv file, click the "All…" menu. Assign the desired name and path in the "Export as" window. Here, the "nominal_values_freq_resp.csv" name is selected.

You can open the saved file in Notepad. As shown in Fig. 2.80, the saved file has three columns. The first column is the frequency in Hertz, the second column is the magnitude in dB, and finally, the third one is the phase in Radians.

Clear the first line before importing the file into the MATLAB®.

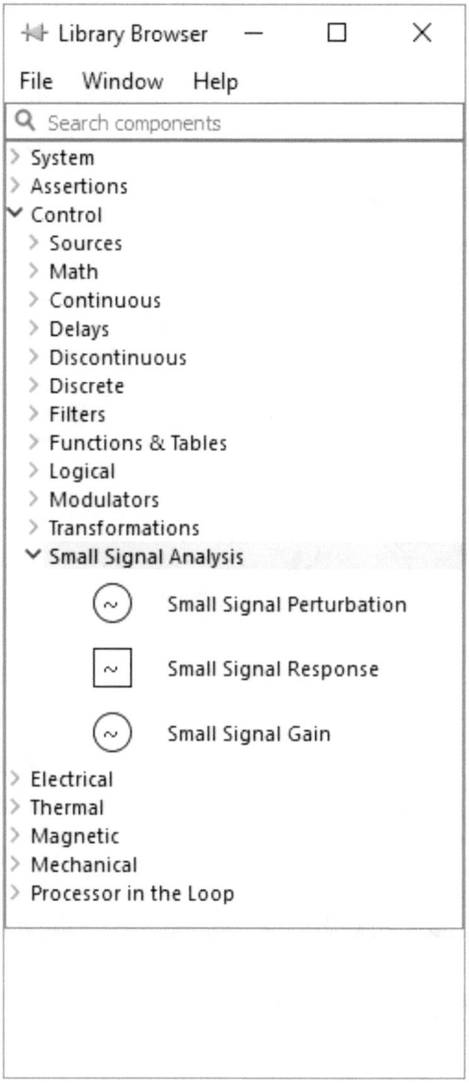

Figure 2.71: "Small Signal Perturbation" and "Small Signal Response" blocks.

Figure 2.72: Converter component values.

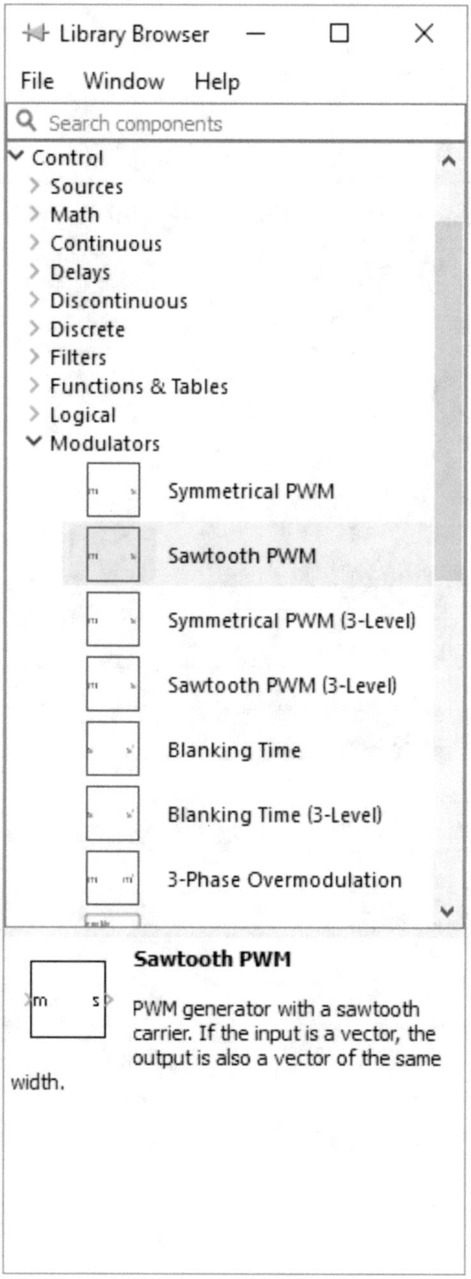

Figure 2.73: "Sawtooth PWM" block.

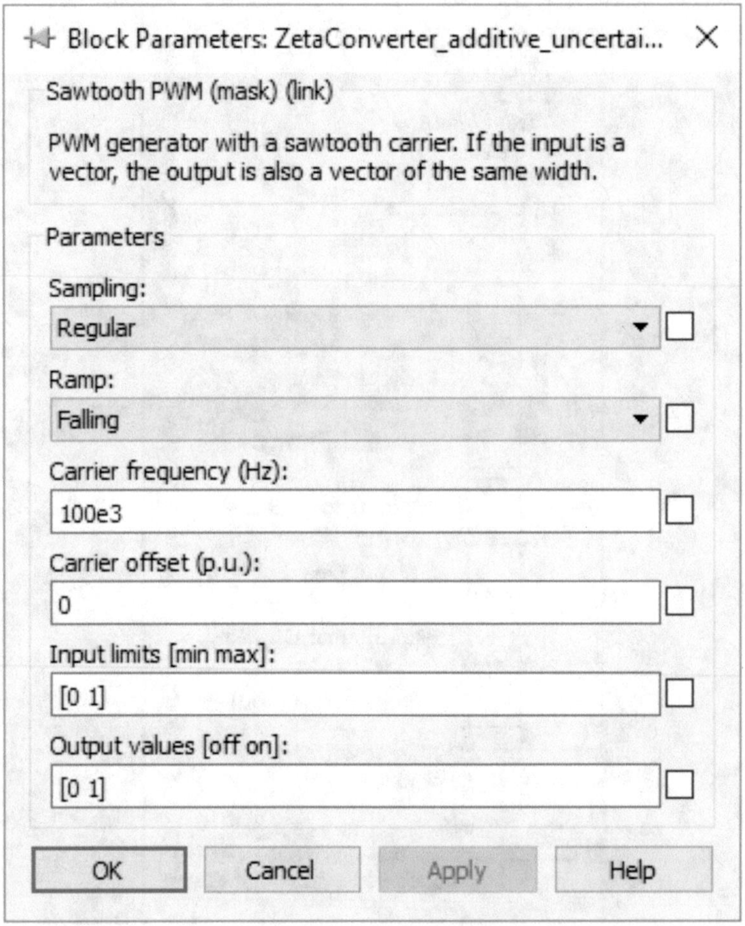

Figure 2.74: Setting of the "Sawtooth PWM" block used in Fig. 2.70.

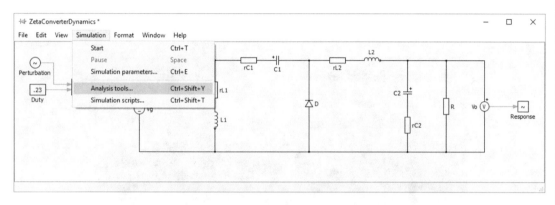

Figure 2.75: Click "Analysis tools…" in order to open the "Analysis Tools" window.

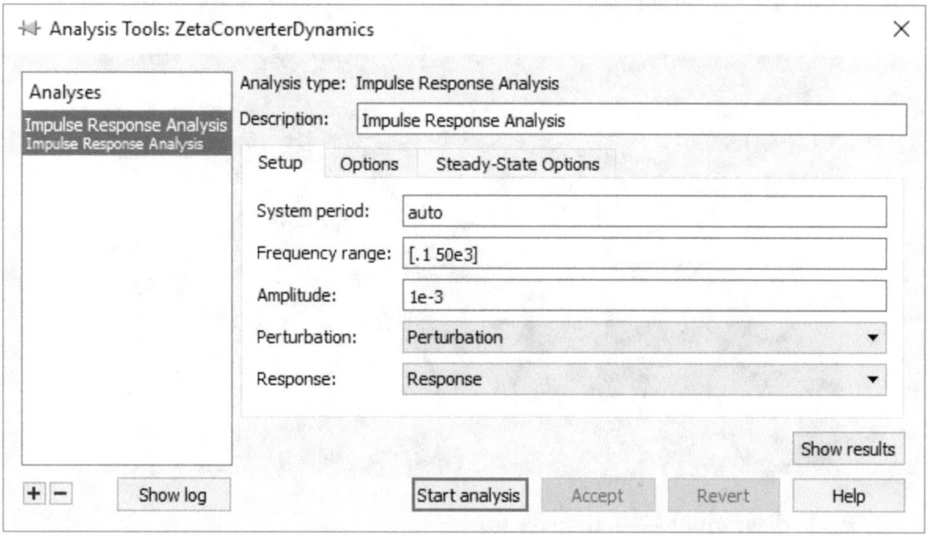

Figure 2.76: The desired frequency range is entered into the "Frequency range" box. The "Amplitude" box is filled with a small number.

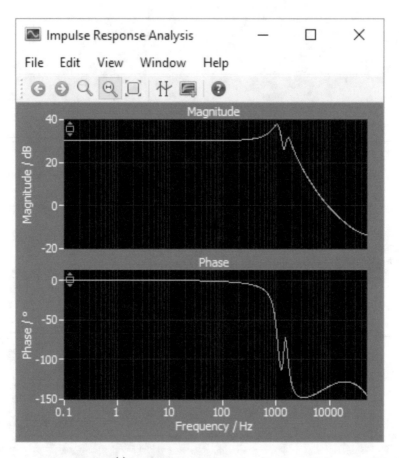

Figure 2.77: Bode diagram of $\frac{v_o(s)}{d(s)}$ transfer function.

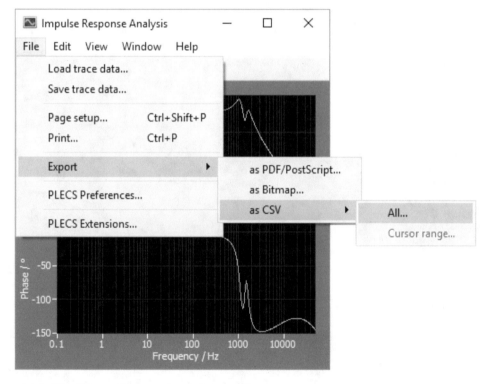

Figure 2.78: Use "Export" to produce the .csv files.

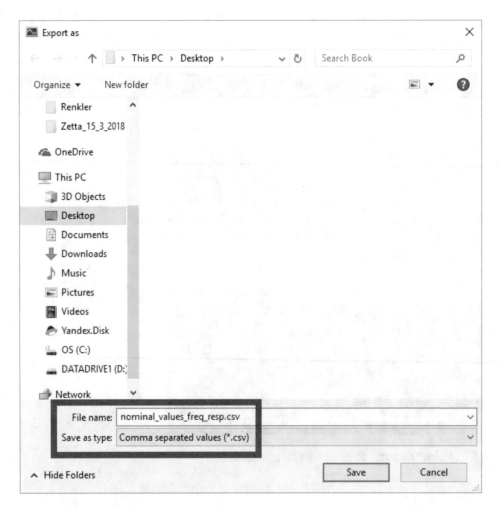

Figure 2.79: The "Export as" window.

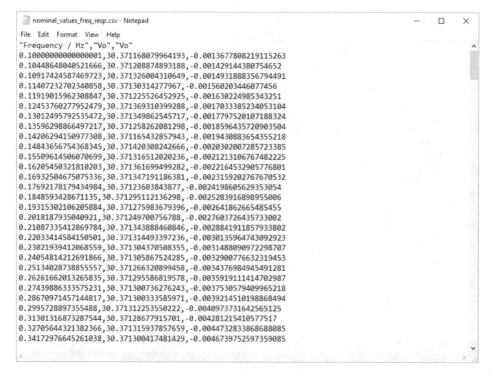

Figure 2.80: The .csv is opened in Notepad.

The following code (Fig. 2.82) reads the .csv file and redraws the Bode diagram in the MATLAB® environment. The drawn Bode diagram is shown in Fig. 2.83. This figure is the same as Fig. 2.77. So, we import the results into the MATLAB® successfully. Using the "tfest" command one can estimate a transfer function for the diagram provided by PLECS®.

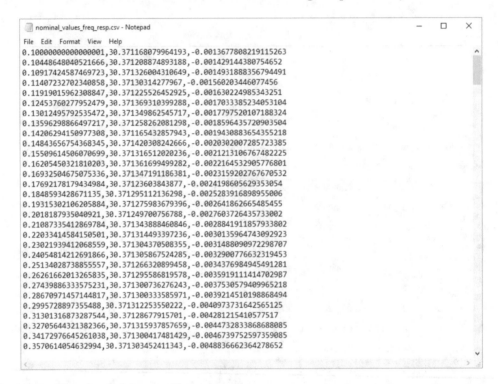

Figure 2.81: First line of the .csv file is cleared.

```
Command Window
>> x=csvread('C:\Users\Dekanlik03\Desktop\nominal_values_freq_resp.csv');%reads the csv file
>> f=x(:,1); % f contains the frequency vector(Hz)
>> w=2*pi*f; % w contains the frequency vector(Rad/s)
>> M=10.^(x(:,2)/20).*exp(j*x(:,3)*pi/180); %frequency response complex form
>> vo_d_nominal=frd(M,w); %nominal system frequency response(frd object)
>> bode(vo_d_nominal),grid on
fx >> |
```

Figure 2.82: Importing the produced .csv file into MATLAB®.

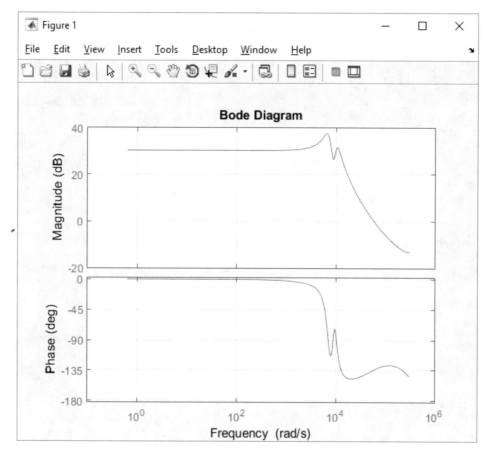

Figure 2.83: Figure 2.77 is redrawn in the MATLAB® environment.

You can use the inductor currents as output as well. For instance, the schematic shown in Fig. 2.84 extracts the $\frac{i_{L_2}(s)}{d(s)}$ transfer function. The "Am1" is an ammeter block. Simulation results are shown in Fig. 2.85.

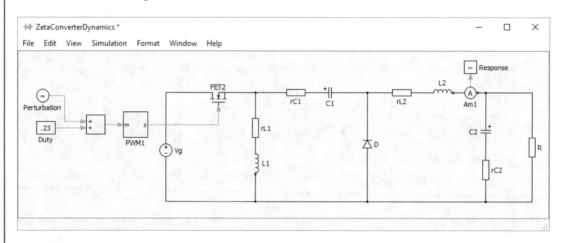

Figure 2.84: Extraction of the $\frac{i_{L_2}(s)}{d(s)}$ transfer function Bode diagram.

You can add the cursor to the shown Bode diagram using the icon.

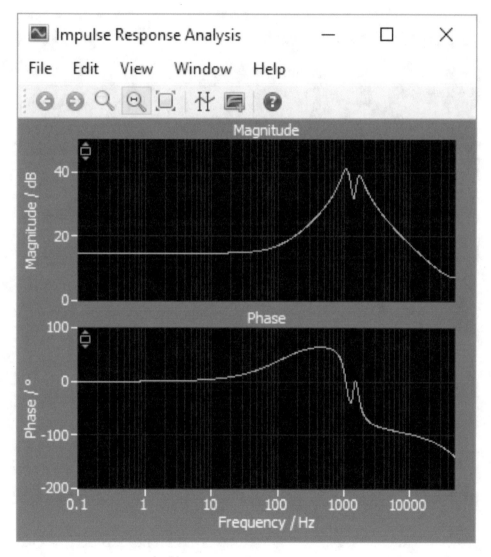

Figure 2.85: Bode diagram of $\frac{i_{L_2}(s)}{d(s)}$ transfer function.

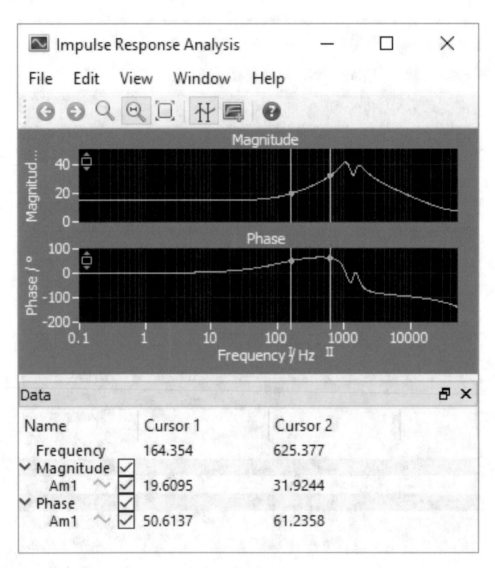

Figure 2.86: Cursors can be used to read the diagram easily.

2.15   AUDIO SUSCEPTIBILITY

The open-loop line-to-output transfer function—also termed power supply ripple rejection (PSRR) or audio susceptibility—is defined as the transfer function from perturbation of the input voltage to perturbation of the output voltage with duty ratio held constant. The audio susceptibility of the studied Zeta converter can be obtained with the aid of schematic shown in Fig. 2.87.

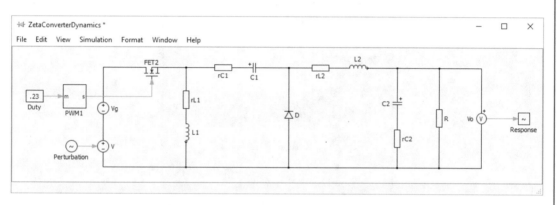

Figure 2.87: Extraction of the $\frac{v_o(s)}{v_{in}(s)}$ transfer function Bode diagram.

The block named "V" in the schematic shown in Fig. 2.87 is a controlled voltage source. After running the simulation with the settings shown in Fig. 2.89, the result show in the Fig. 2.90 are obtained.

2.16   OUTPUT IMPEDANCE

The converter output impedance can be extraceted with the aid of schematic shown in Fig. 2.91. The block named "I" in the schematic shown in Fig. 2.91 is a controlled current source.

After running the simulation with the settings shown in Fig. 2.93, the results shown in the Fig. 2.94 are obtained.

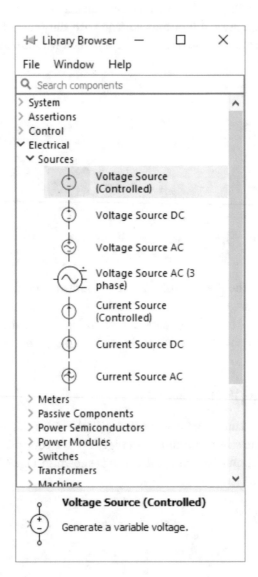

Figure 2.88: Controlled voltage source block.

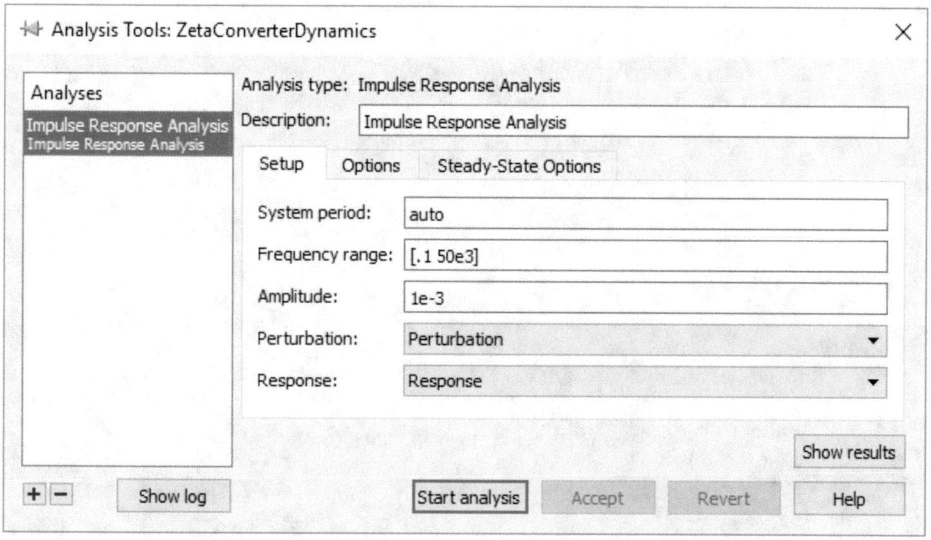

Figure 2.89: Simulation settings.

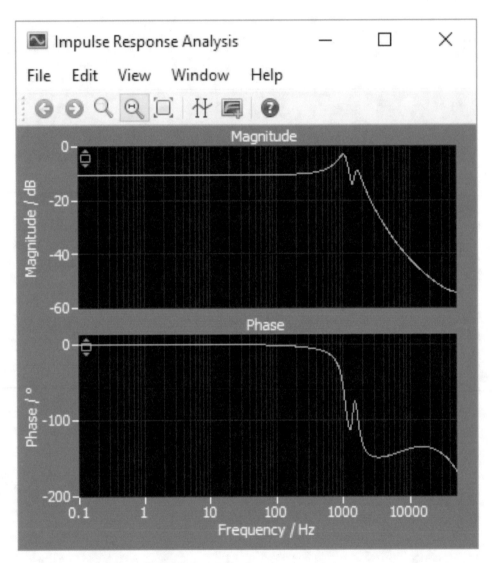

Figure 2.90: Bode diagram of $\frac{v_o(s)}{v_{\text{in}}(s)}$ transfer function.

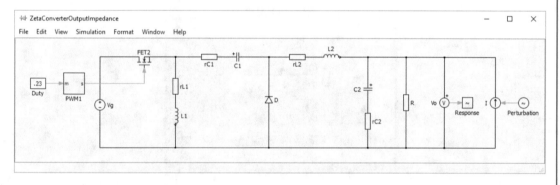

Figure 2.91: Extraction of output impedance.

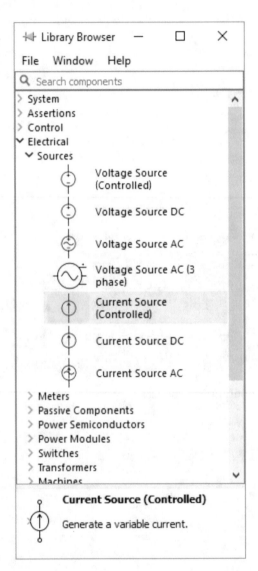

Figure 2.92: Controlled current source block.

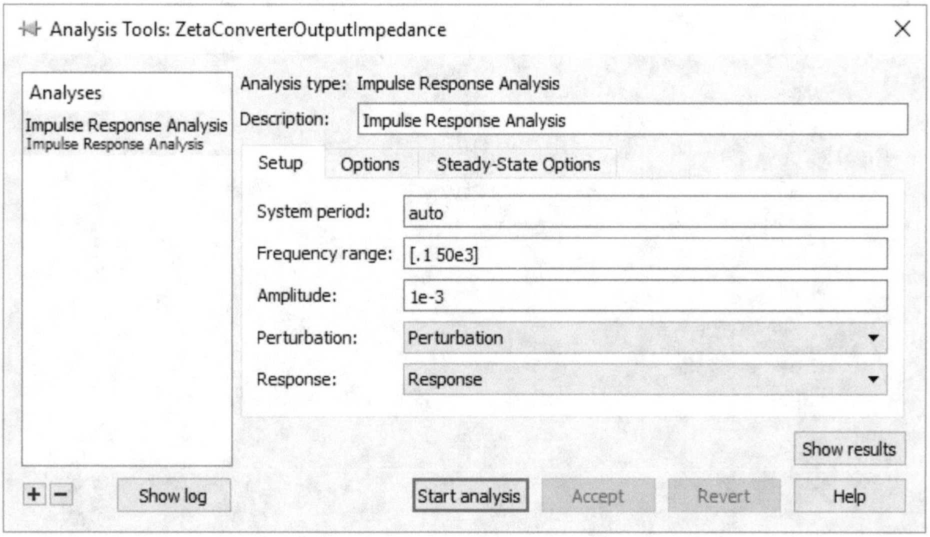

Figure 2.93: **Simulation settings.**

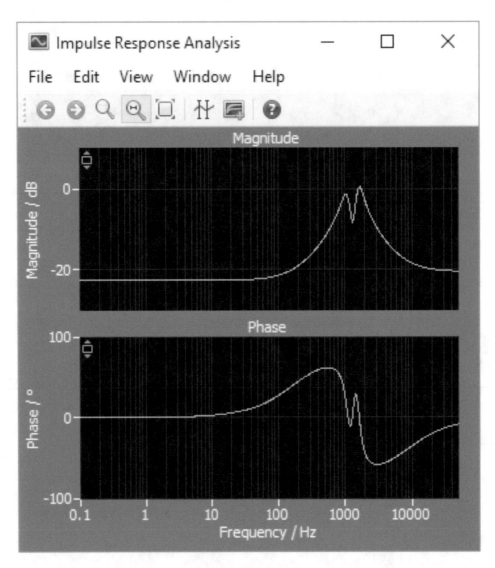

Figure 2.94: Bode diagram of output impedance.

2.17 USING THE PLECS® TO EXTRACT THE UNCERTAIN MODEL OF THE DC-DC CONVERTERS

This section shows how PLECS® can be used to extract the uncertain model of DC-DC converters.

2.17.1 ADDITIVE UNCERTAINTY MODEL

The schematic shown in Fig. 2.95 can be used to extract the additive uncertainty model of the Zeta converter. This schematic is used to extract the uncertain model of $\frac{v_o(s)}{d(s)}$ transfer function.

The schematic is composed of two Zeta converters. The lower Zeta converter has the nominal components. The upper Zeta converter is supplied with the components selected randomly from their allowed interval.

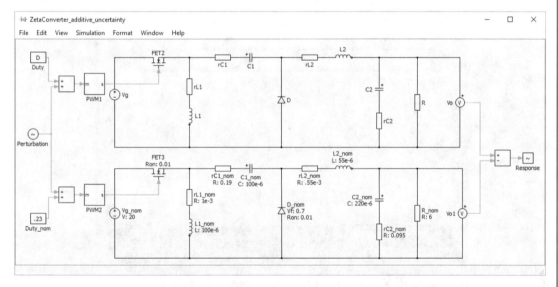

Figure 2.95: Schematic to extract the addtive uncertainty model of the studied Zeta converter.

The components values of the upper Zeta converter is set with the aid of variable names. These variables are initialized in the "Initialization" tab of "Simulation Parameters" window.

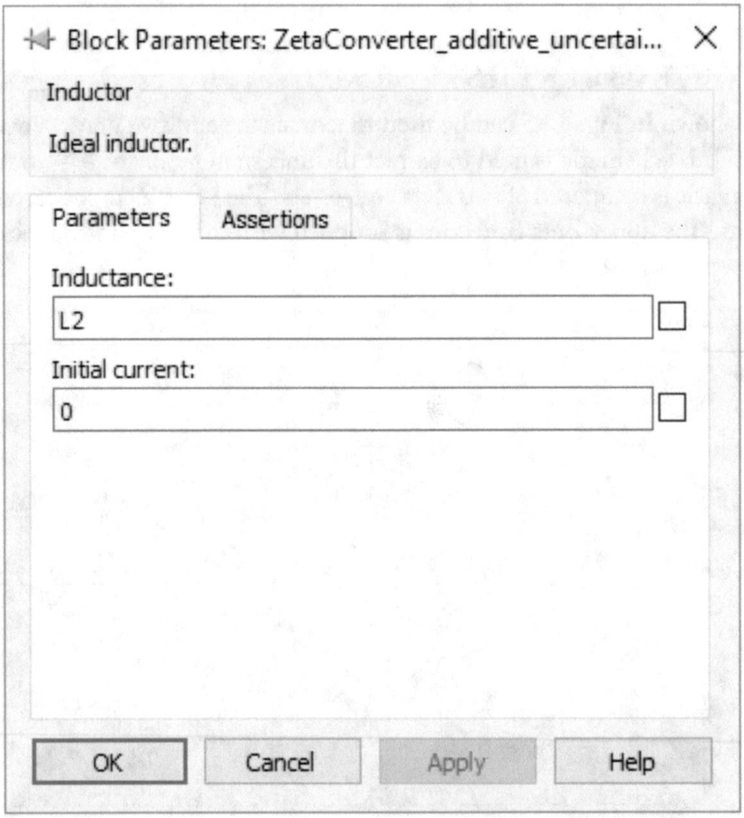

Figure 2.96: The "Inductance" textbox is filed with a variable name.

The components values of the lower Zeta converter is set with the aid of numbers instead of variable names.

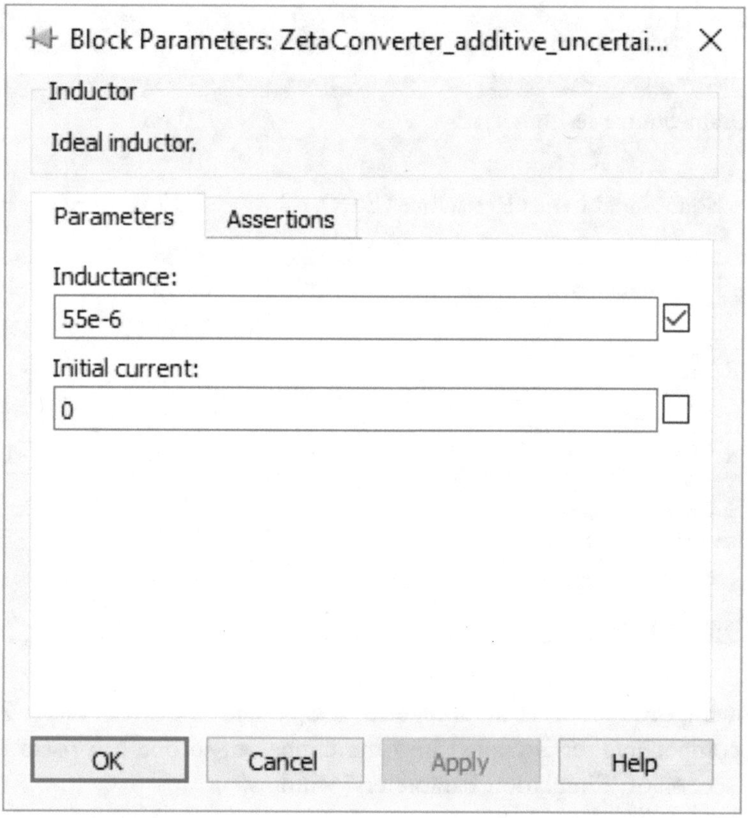

Figure 2.97: The "Inductance" textbox is filed with a numeric value.

Consider a Zeta converter with the following components values and variations; see Table 2.4.

Table 2.4: The Zeta converter parameters (see Fig. 2.1)

	Nominal Value	Variations
Input DC source voltage, Vg	20 V	±20%
MOSFET Drain-Source resistance, rds	10 mΩ	±20%
Capacitor, C_1	100 μF	±20%
Capacitor C_1 Equivaluent Series Resistance (ESR), rC	0.19 Ω	[-10%,+90%]
Capacitor, C_2	220 μF	±20%
Capacitor C_2 Equivaluent Series Resistance (ESR),r C	0.095 Ω	[-10%,+90%]
Inductor, L_1	100 μH	±10%
Inductor ESR, rL_1	1 mΩ	[-10%,+90%]
Inductor, L_2	55 μH	±10%
Inductor ESR, rL_2	0.55 mΩ	[-10%,+90%]
Diode voltage drop, vD	0.7 V	±30%
Diode forward resistance, rD	10 mΩ	[-10%,+50%]
Load resistor, R	6 Ω	±80%
Switching Frequency, Fsw	100 KHz	-

The following code produces an acceptable component set for the upper Zeta converter. The produced components set is copied into the clipboard, so one can paste it easily in the "Initialization" section of "Simulation Parameters" window.

```
%This program produces random component values
%produced values are copied into the Windows clipboard
%so you can paste them easily into the Initialization
%section of PLECS
clc
clear all

NumberOfIteration=20;
    %set the desired number of iteration here.
DesiredOutputVoltage=5;
    %set the desired output voltage here.
```

```
n=0;
for i=1:NumberOfIteration
n=n+1;
%Definition of uncertainity in parameters
VG_unc=ureal('VG_unc',20,'Percentage',[-20 +20]);
   %Value of input DC source is in the range of 16..24
rg=0;
   %Internal resistance of input DC source
rds_unc=ureal('rds_unc',.01,'Percentage',[-20 +20]);
   %MOSFET on resistance
C1_unc=ureal('C1_unc',100e-6,'Percentage',[-20 +20]);
   %Capacitor C1 value
C2_unc=ureal('C2_unc',220e-6,'Percentage',[-20 +20]);
   %Capacitor C2 value
rC1_unc=ureal('rC1_unc',.19,'Percentage',[-10 +90]);
   %Capacitor C1 Equivalent Series Resistance(ESR)
rC2_unc=ureal('rC2_unc',.095,'Percentage',[-10 +90]);
   %Capacitor C2 Equivalent Series Resistance(ESR)
L1_unc=ureal('L1_unc',100e-6,'Percentage',[-10 +10]);
   %Inductor L1 value
L2_unc=ureal('L2_unc',55e-6,'Percentage',[-10 +10]);
   %Inductor L2 value
rL1_unc=ureal('rL1_unc',1e-3,'Percentage',[-10 +90]);
   %Inductor L1 Equivalent Series Resistance(ESR)
rL2_unc=ureal('rL2_unc',.55e-3,'Percentage',[-10 +90]);
   %Inductor L2 Equivalent Series Resistance(ESR)
rD_unc=ureal('rD_unc',.01,'Percentage',[-10 +50]);
   %Diode series resistance
VD_unc=ureal('VD_unc',.7,'Percentage',[-30 +30]);
   %Diode voltage drop
R_unc=ureal('R_unc',6,'Percentage',[-80 +80]);
   %Load resistance
IO=0;
   %Average value of output current source
fsw=100e3;
   %Switching frequency

%Sampling the uncertain set
%for instance usample(VG_unc,1) takes one sample of uncertain
```

```
%parameter VG_unc

VG=usample(VG_unc,1);
   %Sampled value of input DC source
rds=usample(rds_unc,1);
   %Sampled MOSFET on resistance
C1=usample(C1_unc,1);
   %Sampled capacitor C1 value
C2=usample(C2_unc,1);
   %Sampled capacitor C2 value
rC1=usample(rC1_unc,1);
   %Sampled capacitor C1 Equivalent Series Resistance(ESR)
rC2=usample(rC2_unc,1);
   %Sampled capacitor C2 Equivalent Series Resistance(ESR)
L1=usample(L1_unc,1);
   %Sampled inductor L1 value
L2=usample(L2_unc,1);
   %Sampled inductor L2 value
rL1=usample(rL1_unc,1);
   %Sampled inductor L1 Equivalent Series Resistance(ESR)
rL2=usample(rL2_unc,1);
   %Sampled inductor L2 Equivalent Series Resistance(ESR)
rD=usample(rD_unc,1);
   %Sampled diode series resistance
VD=usample(VD_unc,1);
   %Sampled diode voltage drop
R=usample(R_unc,1);
   %Sampled load resistance

%output voltage of an IDEAL(i.e. no losses) Zeta converter
%operating in CCM is given by:
%         D
%VO=--------VG
%        1-D
%where
%VO: average value of output voltage
%D: Duty Ratio
%VG: Input DC voltage
%So, for a IDEAL converter
```

```
%         VO
%D=----------
%      VO+VG
%Since our converter has losses we use a bigger duty ratio,
%for instance:
%            VO
%D=1.1 ----------
%          VO+VG
D=1.1*DesiredOutputVoltage/(VG+DesiredOutputVoltage);

%preparing the strings
S1=strcat('VG=',num2str(VG),';');
S2=strcat('rds=',num2str(rds),';');
S3=strcat('C1=',num2str(C1),';');
S4=strcat('C2=',num2str(C2),';');
S5=strcat('rC1=',num2str(rC1),';');
S6=strcat('rC2=',num2str(rC2),';');
S7=strcat('L1=',num2str(L1),';');
S8=strcat('L2=',num2str(L2),';');
S9=strcat('rL1=',num2str(rL1),';');
S10=strcat('rL2=',num2str(rL2),';');
S11=strcat('rD=',num2str(rD),';');
S12=strcat('VD=',num2str(VD),';');
S13=strcat('R=',num2str(R),';');
S14=strcat('D=',num2str(D),';');

%coping the data into the Windows Clipboard.
%So, you can paste it into the PLECS
data=strcat(S1,S2,S3,S4,S5,S6,S7,S8,S9,S10,S11,S12,S13,S14);
clipboard('copy',data)
disp('Data is copied into clipboard. You can paste it in
   PLECS initialization section right now...')
message=strcat('Iteration #',num2str(n),' finished.');
disp(message)
disp('Press any key to produce another value set.')
disp(' ')
pause
end
```

```
disp('-------------------------')
disp('Program terminates here...')
```

After running the code (and finishing the first iteration), the message shown in Fig. 2.98 will appear.

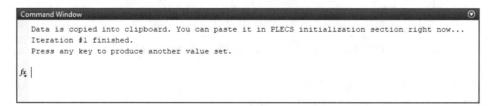

Figure 2.98: User is asked to press a key before the next iteration.

Before pressing any key, the user goes to the "Initialization" tab of "Simulation Parameters" window and paste the produced data (Fig. 2.99).

Now the user runs the PLECS®simulation with the values produced by the MATLAB®code.

The user exports the obtained result as an .csv file (see Fig. 2.102).

After saving the result, the user retuns to the MATLAB®environment and press a key in order to produce a new component set. The user redoes the above-mentioned process until all the iterations are finished. For instance, the number of iteration is set to 20. So, the MATLAB®code produces 20 different components sets. All the simulation results are exported as .csv file.

All the obtained .csv file are opened in the Notepad and the first line is cleared.

Figure 2.99: The "Initialization" tab of "Simulation Parameters" window.

Figure 2.100: Produced random values are paste into the "Model initialization commands" section of "Initialization" tab.

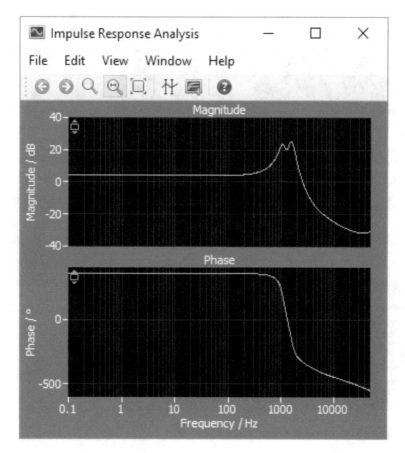

Figure 2.101: Simulation result.

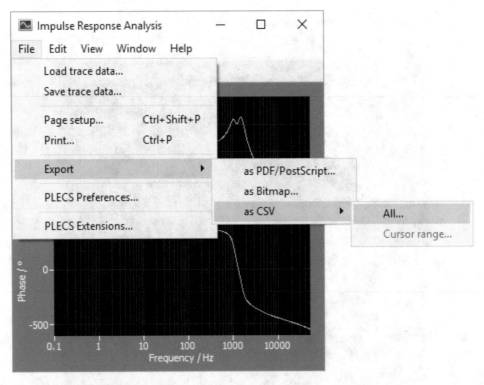

Figure 2.102: Obtained result is exported as a .csv file.

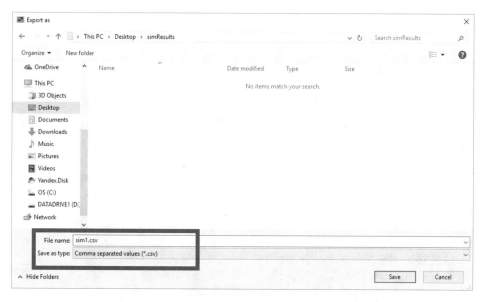

Figure 2.103: The name sim1.csv is selected for the produced .csv file.

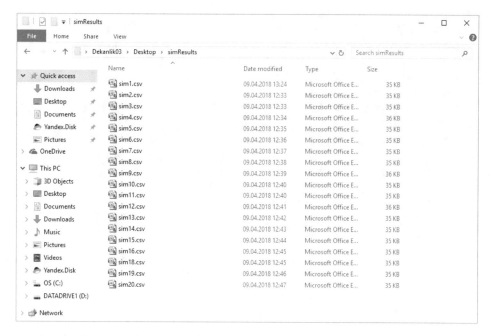

Figure 2.104: Analysis results are saved as sim1.csv, sim2.csv, sim3.csv, ..., sim20.csv.

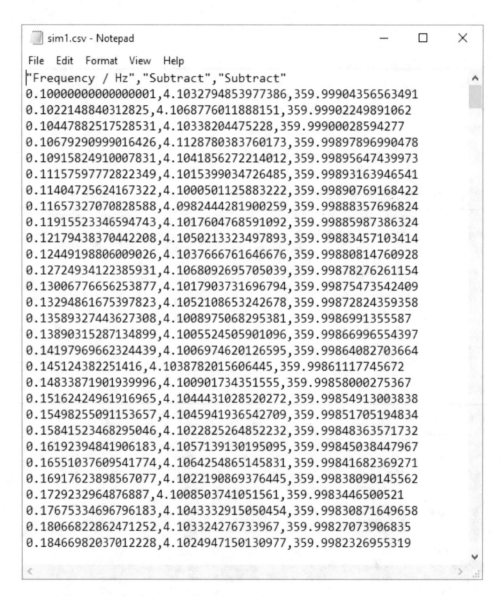

Figure 2.105: Opening the "sim1.csv" in Notepad.

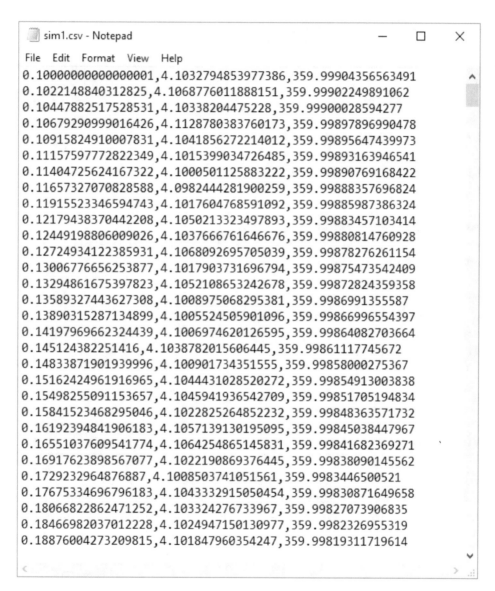

Figure 2.106: First line of sim1.csv is cleared.

The following MATLAB®code reads all the simulation results (prepared .csv files) and draws them on the same graph.

```
%This program draws the results(Additive Uncertainty)
%produced by PLECS
clc
clear all
NumberOfSimulations=20;

for i=1:NumberOfSimulations
disp('percentage of work done:')
disp(i/NumberOfSimulations*100)
disp(' ')

name=strcat('sim',num2str(i),'.csv');
x=csvread(strcat('C:\Users\Dekanlik03\Desktop\SimResults',
   name));    %reads the csv file
f=x(:,1);
w=2*pi*f;
M=10.^(x(:,2)/20).*exp(j*x(:,3)*pi/180);
   %Magnitude of freq. resp.(complex form)
H=frd(M/Mnominal,w);
bode(H), grid on
hold on

end
title('Additive uncertainty in vo(s)/d(s)')
```

After running the code, the result shown in Fig. 2.107 are obtained.

After obtaining the graph shown in Fig. 2.107, the user can use the techniques shown in previous chapters in order to find suitable weight which passes above all the transfer functions.

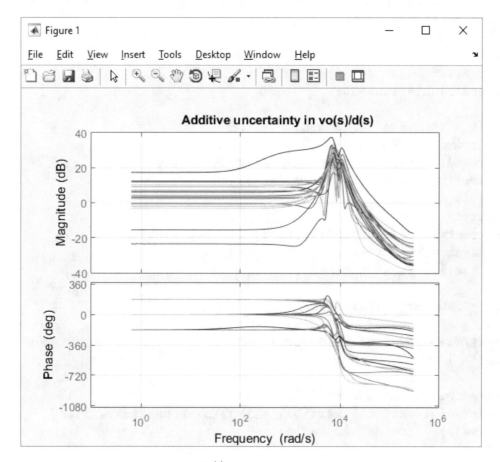

Figure 2.107: Additive uncertainty in $\frac{v_o(s)}{d(s)}$.

2.17.2 MULTIPLICATIVE UNCERTAINTY MODEL

The schematic shown in Fig. 2.95 can be used to extract the additive uncertainty model of the converter. The following code normalizes (divides the difference to the nominal transfer function) the obtained results. This lets the calculation of multiplicative uncertainty weights.

```
%This program draws the results(Multiplicative Uncertainty)
%produced by PLECS
%Running this code takes time, please be paitient!
clc
clear all
NumberOfSimulations=20;

%Results obtained for nominal system
%see previous analyses
x=csvread(strcat('C:\Users\Dekanlik03\Desktop
    \nominal_values_freq_resp.csv'));
f=x(:,1);
w=2*pi*f;
M_nominal=10.^(x(:,2)/20).*exp(j*x(:,3)*pi/180);
    %Magnitude of freq. resp.(complex form)
vo_d_nominal=frd(M_nominal,w);

for i=1:NumberOfSimulations
disp('percentage of work done:')
disp(i/NumberOfSimulations*100)
disp(' ')

name=strcat('sim',num2str(i),'.csv');
x=csvread(strcat('C:\Users\Dekanlik03\Desktop\SimResults',
    name));    %reads the csv file
f=x(:,1);
w=2*pi*f;
M=10.^(x(:,2)/20).*exp(j*x(:,3)*pi/180);
    %frequency response complex form
H=frd(M/M_nominal,w);
bode(H), grid on
hold on
end
title('Multiplicative uncertainty in plant')
```

After running the code, the result shown in Fig. 2.108 is obtained.

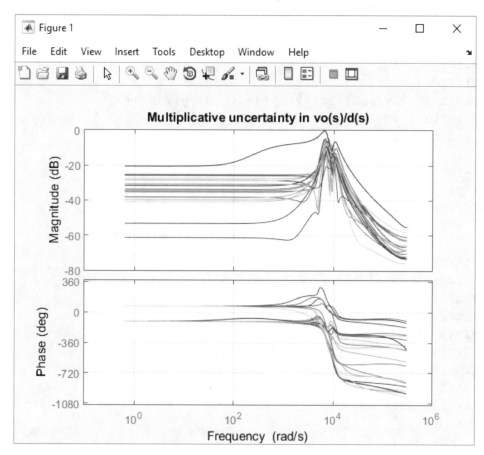

Figure 2.108: Multiplicative uncertainty in $\frac{v_o(s)}{d(s)}$.

An upper bound for the transfer functions shown in Fig. 2.108 can be found with the aid of the following code. Fitted weight is shown in Fig. 2.109.

```
%This program draws the results(Multiplicative Uncertainty)
%produced by PLECS
%You can obtain the upper bound of uncertainty as well.
%Running this code takes time, please be paitient!
clc
clear all
NumberOfSimulations=20;
```

```matlab
%Results obtained for nominal system
%see previous analyses
x=csvread(strcat('C:\Users\Dekanlik03\Desktop
    \nominal_values_freq_resp.csv'));
f=x(:,1);
w=2*pi*f;
M_nominal=10.^(x(:,2)/20).*exp(j*x(:,3)*pi/180);
    %Magnitude of freq. resp.(complex form)
vo_d_nominal=frd(M_nominal,w);

for i=1:NumberOfSimulations
disp('percentage of work done:')
disp(i/NumberOfSimulations*100)
disp(' ')

name=strcat('sim',num2str(i),'.csv');
x=csvread(strcat('C:\Users\Dekanlik03\Desktop\SimResults',
    name));   %reads the csv file
f=x(:,1);
w=2*pi*f;
M=10.^(x(:,2)/20).*exp(j*x(:,3)*pi/180);
    %frequency response complex form
H=frd(M/M_nominal,w);
bodemag(H), grid on
hold on
end
title('Multiplicative uncertainty in plant')

pause
disp('press any key to continue')

%Selection of the upper bound for uncertainty
Number_of_points=20;
[freq,resp_dB]=ginput(Number_of_points);
%Reading are in desibel(dB). The following loop
%find the magnitudes.
for i=1:Number_of_points
    resp(i)=10^(resp_dB(i)/20);
end
```

```
selected_points_frd=frd(resp,freq);
  %Making the frd object.
ord=3;
  %Order of produced weight
W=fitmagfrd(selected_points_frd,ord);
  %Fitting a transfer function to the selected data points.
Wtf=tf(W);
bode(Wtf,'r--')
```

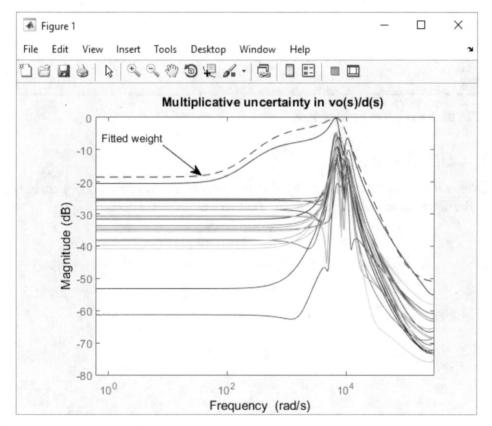

Figure 2.109: Fitting an upper bound to the obtained uncertain set.

One can obtain the calculated weight with the aid of commands shown in Fig. 2.110.

```
Command Window                                                    ⊙

  >> zpk(Wtf)

  ans =

    0.0026377 (s+110) (s^2 + 2.067e05s + 1.517e10)
    -----------------------------------------------
          (s+626.9) (s^2 + 5653s + 5.902e07)

  Continuous-time zero/pole/gain model.

fx >> |
```

Figure 2.110: Equation of fitted weight.

2.18   CONCLUSION

This chapter studied the uncertain model of Zeta converter. Proposed techniques can be used to extract the uncertain models of other types of converters.

Authors' Biographies

FARZIN ASADI

Farzin Asadi is with the Department of Mechatronics Engineering at the Kocaeli University, Kocaeli, Turkey.

Farzin has published 30 international papers and 10 books. He is on the editorial board of 6 scientific journals as well. His research interests include switching converters, control theory, robust control of power electronics converters, and robotics.

SAWAI PONGSWATD

Sawai Pongswatd received a B.E., a M.E., and a D.E. from King Mongkut's Institute of Technology Ladkrabang (KMITL). Currently, he is an associate professor at the KMITL, Bangkok, Thailand. Assoc. Dr. Sawai Pongswatd is a chairman of technical committee (TC69) of the Thai Industrial Standards Institute and an instructor at the Fieldbus Certified Training Program (FCTP). His research focus on power electronics, energy conversion, and industrial applications.

KEI EGUCHI

Kei Eguchi received his B.E., M.E., and D.E. from Kumamoto University, Kumamoto, Japan, in 1994, 1996, and 1999, respectively. Currently, he is a professor at the Fukuoka Institute of Technology, Fukuoka, Japan.

Dr. Kei is a president of the Intelligent Networks and Systems Society, an associate editor of *IJICIC*, an associate editor of *ICIC Express Letters*, and a senior member of IEE of Japan, APCBEES, IRED, and SAISE. He has published more than 200 international papers. His research interests include switching converters, nonlinear dynamical systems, and intelligent circuits and systems.

NGO LAM TRUNG

Ngo Lam Trung received a B.E. and a M.E. in computer engineering from Hanoi University of Science and Technology, Hanoi, Vietnam. He received a Ph.D. in functional control systems from Shibaura Institute of Technology, Tokyo, Japan. His research interests include embedded system, embedded software development, and mobile computing.

Printed in the United States
by Baker & Taylor Publisher Services